THE CONSCIOUSNESS OF THE UNIVERSE

Visualizing the Universe Through a Schizophrenic Mind

SUHAIL AHMAD PARAY

INDIA • SINGAPORE • MALAYSIA

Alim Gaw Alim-e-Ludni

Tchalim Shakh Aam Badni

Porum Na Kaseh Nish Toosh

Karis Aarasteh Yaari

Tunn Naar Daez Aarwali

Kyoh Kalee karu Thehraav.

– Rasool soab

Yaavunn Lolhath loli,

Bei Akki Latti Roi Haav,

Dohay Vuchh Hai Thali Thali,

Kyoh Kali Karu Thehraav (Samad mir[R.A])

Zaatas sipatas chai milwan

Khasul Khass chue manz khassan

Vani deu har zarmaa chus tchen

Khasul Khass chue manz khassan

Sheshpash trupeth shishkal wazan

Ishaar tamkue che husharan. (shamas faqir[R.A])

– Bashir Soab

Contents

Preface

This book right now in your hands has been written and edited keeping in view the various theories and thesis¬- obviously scientific ones, That in one way pave way for scientific thought but on the other hand, create only the doubts that prove to be havoc if remain unaddressed or unnoticed. Thus the sole aim of this book is to address all those facts and realities that comprehend all those truths that's why I have given it the title:-

'The Consciousness of the Universe'

One more reason to give it such a name is: That I have always been under estimated and under mined by my peers, relatives even some of my teachers.

Thanks to my mummy and papa who never doubted my potential.

Last but not the least, my patron, friend and mentor. Mr. Hilal Ahmad Paray who time and again encouraged my worth and keep me going on in formulating and designed this book for you- "consciousness of universe".

One more cause for bestowing my book such a name is that I am actually despised with a kind of mental illness- namely schizophrenia and OCD that gave way to my friends and my relative to make fun of me and often play truants against me but I always kept going with my own world view and I only tried to impress upon them the way I view and overview the universe- infact, we forget the very basic principle that the nature has evolved us not to search any truth rather we are the very DNA carriers of that truth.

So, incase, my dear readers find anything hard to sublimate or understand from this book, I shouldn't be blamed rather the matter should be related and addressed to the nature.

You must have read, learn and understood the laws and rules of physics.

Time

It is an illusion- The present, the past or the future- this all is an illusion and nothing more. The value of the 'amount' is very much prominent and important with respect to the processes of the universe, the increase or the decrease of that 'amount' becomes a valid and basic factor in determining the time- and that gives the 'time' that much essence and that is why we can experience the time. The amount represents a quantity that we call time.

The nature creates and it creates a masterplan of this universe for what has been done or what is going to happen. That all is stored in the form of information which exists in the form of a pattern, a sequence- so, is the universe the same that we have observed it to be? or is it just an illusion yet? For our feeling and thinking is only the foreplay of certain chemicals inside us that take the preliminary mode in deciding a thing in one way or the other- may be the truth is something else and we perceive it as some other thing- is there any place for the human feeling in the process of the universe? Do the rules and laws of universe know about their existence?

Does feeling only alter our decision making or does it affect the very truth as well?

Laws and Rules

It refers to the way through which a scientific process takes place. it is a path from the beginning up to the end. we do know our limits and boundaries, so we often act within those limits or extents or boundaries.

Does science, especially physics, share the same feelings or does it exhibit the same feelings? If yes then the next question will ultimately follow regarding it and that is how?

In scientific language, from beginning up to the end, it goes with different steps and because of those rules and laws it makes us confined within certain limits of what is possible and what is not possible, and if we could discover or penetrate through these rules and laws discovery or any other inventions, we can peep into these rules and laws. since we can penetrate in these rules and laws, then we can even play with these rules and laws.

So ultimately, it can be inferred that science, especially physics, does not share any of such things into knowledge regarding its being, otherwise, it would not have let us penetrate a peep through it. Now with the various scientific discoveries and inventions and yes, with the advent in technology we have out witted even the science in that whoever or whatever follows such scientific method, will always get the same results or outcome always as is exhibited by science itself.

The support from physical phenomenon works in the same fashion, it always exhibits the same work functions until we change some of the external factors related to it; these are the same factors that we change, that bestow us with the results, that we call as the inventions so by changing some of or all the external factors regarding any physical phenomenon, we change the course of

The whole phenomenon or we can say that we change.

The directions of it being and that is all.

Let's take an example of four stroke internal combustion engine.

When we fuel it a spark comes out of the ignition coil which eventually turns into a blast like situation that push down the piston, which in-turn results in the rotatory motion of the shaft, the fuel just goes on to burn only to provide energy it has nothing to do with the rotation of the shaft, rather it combusts when comes into contact with oxygen only to provide the desired energy that gets released into form of heat and light. it has nothing to do with the working of the whole engine. It does what it is confined to do.

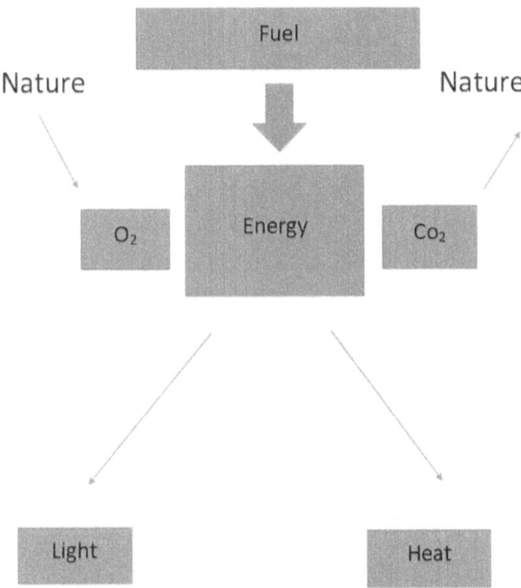

The diagram that is more like a flow chart shows the working principle of how fuel comes into contact with the oxygen that leads to its combustion releasing energy in the form of heat and light and some of the exhaust like co_2 etc.

Nuclear Fusion

It simply refers to the process by which an unstable nucleus loses energy by radiations.

Nuclear Fission

It is the very process through which the energy comes out of the radiating nucleus but its sole aim isn't, and never was, to give rise to destruction nuclear warhead, rather it is the very human effort, in terms of interfering its course of nature, that has shaped it into such destructive warheads as we know it today.

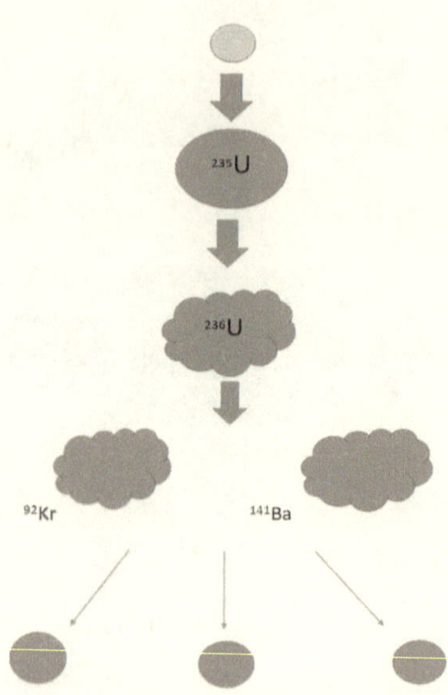

The laws adhered by the path may well be regarded as the sense. the very sense with which it can detect what to do next and how to do it.

Previously we have discussed that weather physics has any intensions about it being or not.

If it knew its presence, then all those discoveries and inventions that have been possible till date would have been all together impossible. Because it wouldn't have allowed to alter its own rules, that to affect its own being.

The physical laws are blind but with same extra high sense, that is it know what to do the next. let's take example of gravity to illustrate this. My view point

Cause of Gravitation and Its Graphical Analysis

Suhail Ahmad paray

Islamic. University. Of science and technology

Abstract: In this paper I am preparing a mathematical model for the cause of gravitation and its graphical analysis. This shows the relationship between the density, kinetic energy and gravitation.

Key words: Gravitation, density, energy equilibrium, kinetic energy and energy exchange.

1. Introduction

Since all planets, the sun, even the galaxy, we belong to and the other galaxies as well are in constant motion with respect to each other and with respect to the universe not only this, rather, universe as a whole has some definite speed of expansion. Now taking the case of our planet i.e., the earth.it has a speed of 30km/s in its orbit.and with respect to solar system its speed has been found some 2okm/s and with respect to galaxy its speed is detected some 400_600km/s and with respect the universe as a whole its speed is 2.1_74.2km. Since the earth as a whole moves round its axis.

the core itself, through in a molten state is also going through with it. Thus every atom thereof wants to attain the stable state at the very central position of the core. Because the instability of energy is created towards the end position. since the atoms at the very central position attain a stable state of energy are pushed to the ends by those atoms that have energy instability so that they became stable at the central position.And as such, these atoms that would have got stability at the centre and are pushed toward the ends getting an energy instability, produce such an effect to attain energy from other atoms so that to retain their energy levels... (((I.e. A complete energy exchange takes place))). And this is a hiding principle of gravitation. Now this effect produced depends upon……

The size of centre of celestial body.

Velocity of the centre revolving.

Size of the Centre

Grater the size of the centre means greater will be the number of atoms producing such effect. The centre is, in fact, 1/5.2279508197.the part of the radius of the planet or the celestial body.

Speed of the Centre/Celestial Body

Greater the speed of the centre/celestial body greater will be effect produced. because the atoms get unstable at a tremendous rate and same will be effect produced by them on other atoms outside the centre they are associated to. The velocity of the centre in fact depends upon the density of the matter outside it. BECAUSE.

Different densities have different energy levels or in other words density has a specific energy level. SINCE,

$E = MC^2$—(1) theory of Relativity

I am in search of this.........

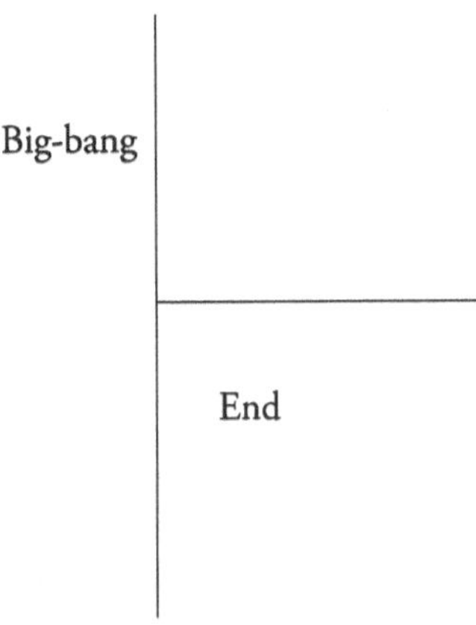

But $D = M/V$___(1)

Density = Mass/Volume

Mass = Density × Volume_________(2) putting value of M from (2) in (1)

$E = DVC^2$

I.e. E directly proportional D and E directly proportional V [:C^2 is constant of proportionality]

Now E Directly Proportional D

Treating E as K.E, we can write, K.E of the centre and hence the moving particles thereof is directly proportional to the Density of celestial body. Since the density has its own specific energy level and as such the K.E of the centre correspond to the energy level of the density of the matter outside it. Why so: the density outside the centre corresponds to the K.E that suits and is in proportion to the energy level of the density concerned. And conversely, the K.E of the particles in the centre produced their effect proportion to the K.E of the particles in the outer centre matter. Since the density of the particles outside the centre is high and thus their K.E is always high In comparison to K.E of the particles of the earth. Thus to attain equilibrium state between the energy levels of the centre and the outer centre particles, they produce an effect in according with the K.E of particles outside the centre and thus a sure cause of gravitation as well...

2.1 Graphical Analysis

Graphical representation of the K.E /speed acquired by the particles of the centre VS the density of the surface particles given below:

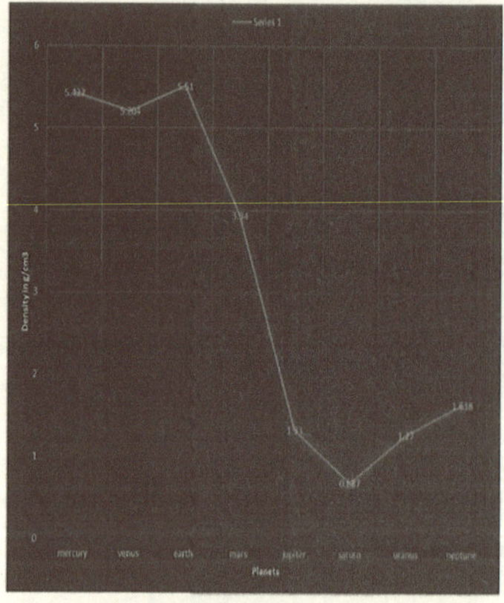

Shows that the two graphs are similar with each other hence supporting our view......

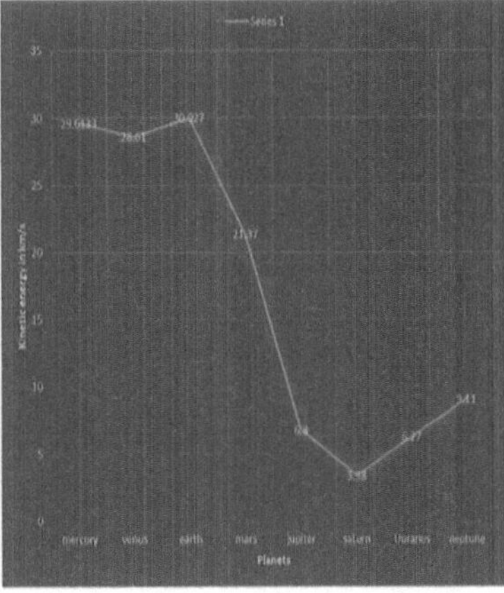

Now the question arises can we change the gravitational pull or gravitational effect produced by the central particles of a body? And if so, How?

The answer to this question is very easy oriented by that too hypothetically. Suppose the surface density is increased, in turn the particles of the centre get effected and their K.E also rises and in turn we have a strong gravitational pull.Else if energy level of surface particles changes when and if... They move on their own freely that two results in increase in the gravitational effect produced the centre particles......

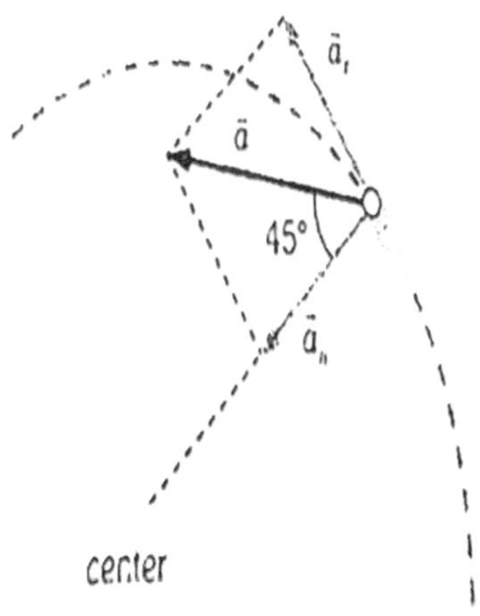

Figure showing the particles of the surface moving freely.

2.2 If I formulate a formula

Here is a formula that gives us the appropriate value of gravitational force exerted by any celestial body. The formula is.......

$F \propto Rc$___(1) [F force of gravitation, Rc Radius of centre]

$F \propto V$____(ii)[velocity of centre of body]

$F \propto RcV$

$$F = \frac{RCV}{SP}$$

Where "sp" is constant of proportionality.

From the equation it is clear that...

Sp=RcV/F _____[and it's units depend on the units of radius, velocity and force.....and its value has been determined to be 3734693877.551 ms^{-1}/Kg. It is clear from the equation that units of force are "N"....

As units of F, R and velocity are Newton, meter and m/s respectively.

$$F = \frac{R_C V}{SP}$$

$$SP = \frac{RCV}{F}$$

$$= \frac{1m \times 1m/s}{1N}$$

$$= \frac{1m^2}{N.S}$$

$$= \frac{1m^2}{1kg \frac{m}{S^2} \times 1s}$$

$$= \frac{1m \times s}{1kg}$$

$$= 1ms/kg$$

Now putting this value in given=n

$$F = \frac{1m}{1ms/kg} \times 1m/s$$

$$= 1kg\ m/s^2$$

Hence justified......

Energy exchange relation between energy exchange and gravitational waves;

When energy exchange between particles they create an effect on each other which is an attractive force in the form of waves. These waves are called gravitational waves and such waves are created due to the exchange of energy between two particles.

1. Greater the effect of gravitational waves greater will be the size of center.

2. **Speed of the canter /celestial body;**

Greater the speed of centre greater will be the effect produced by them on other atoms outside the centre they are associated to. The velocity of centre in fact depends upon the density outside it.

Relation between kinetic energy and surface density in celestial bodies;

$E = MC^2$________(1) theory of Relativity But $D = M/V$______I.e Density=Mass/Volume.

Mass = Density Volume________(2) putting value of M from (2) in (1)

$E = DVC^2$

I.e. E directly proportional D and E directly proportional V [:C² is constant of proportionality]

When celestial body attains energy which can easily superimpose with energy related to surface density and creates share in formation of gravitational waves. Through the mechanism of energy exchange between the particles in the centre

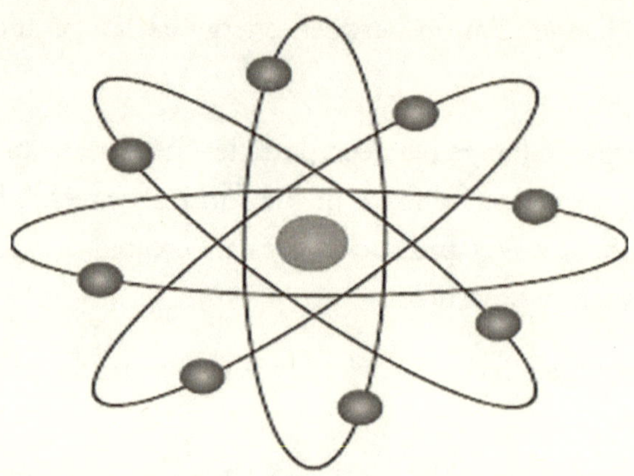

Force of Gravitation at Atomic Level

Ÿ In atoms we saw, electrons are constantly moving round the nucleus in their concerned orbits. And the nucleus, you know, consists of protons which are positively charged particles.And the position change of these protons is responsible to set the negatively charged electrons in a circular motion round the nucleus. Not only this, the protons, themselves, get initiated due to the force exerted by the electrons over them, in turn Quite ultimately, every electron as well as the protons tries to attain a more stable position.OR in other words, that position where their respective energy level gets more stabilized. Since that position is core or the centre or the nucleus, itself. But the electron can't enter that centre because of their orbital energy binds then there at their concerned shells. So,

It is possible only for protons. And because of the moving phenomenon of the protons, every proton tries to get the innermost more stable, position, in the nucleus.And as that is attained, the other protons that are trying to do so it towards the border of the nucleus. The stable position with least energy to the position where its energy is quit unstable, results in an effect that we all known to be gravitational force.

The factor that determines this gravitational force here are............

I. The size of nucleus especially it's radius.?. *II.* Stable high energy level......

The Radius: Greater the size or greater the size of nucleus, Greater will be the no of protons producing such effect. In an atom as a whole......

Ÿ Stable high energy level:- Greater the energy levels of shells, Greater will be the effect produced by these on the protons moving inside the nucleus and in turn, Greater will be the difference between the stable and unstable energy levels of the protons and hence greater will be the gravitation effect produced by them......

Since electrons are negatively charged particles and they induce an attractions pull on the positively charged protons only i.e. only protons in the nucleus are moving and not the neutrons as they don't share any charge!!!!......

Black holes;

What happen when black hole originates. The surface density of a dying star increases and attains high energy level and the gravitational effect increase

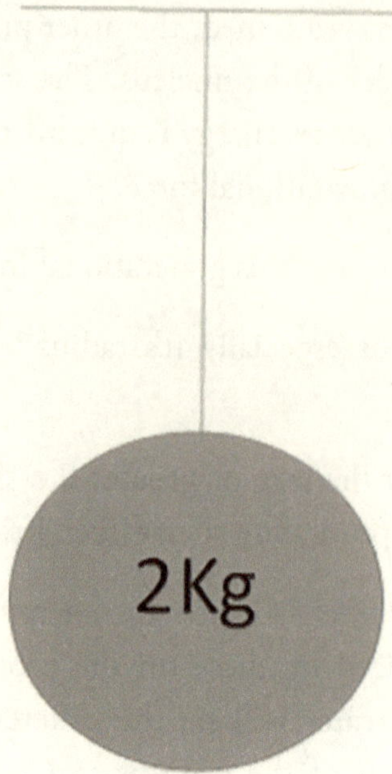

let›s take example of law of gravitation.

First we have to unfold cause of gravity, if we take the example the relativity theory, it tells us that the gravity is not any force but as the consequence of the curvature of space-time caused by un even distribution of mass in Einestines view, gravity is the curvature of space time, caused by massive objects, in which huge mass results in the creation of huge gravitation field because of high curvature of space and time.

now the mass of 2kg, that we have just tied to the hook with the help of a string the force that acts on it can be measured using newton's method I,e.

F = GMm

R^2

Where

M ⟶ mass of earth

m ⟶ Mass of the objects

R ⟶ The radius of the earth

Now what is the mass? And how apart are the two bodies, these factors will be determined the magnitude of force that exits between the two bodies.

The path followed by this whole process, is what I call the sense of physics.

We fool the same sense and the ultimate outcome is the inventions we simple change the path with regard to the physical laws.

Natural Selection and Physics

It is the key mechanism of evolution of the universe. I have already discussed the fact that laws and rules of physics are just blind. Now let's talk of natural selection, it too moves along with the blind sense but with extra high-tech sense and it is because of the same natural selection that the universe came into existence that, two, out of nothing.

Thought on Origin of Universe my View Point

Over the centuries, we have viewed the universe through many different lenses, our models and theories have shifted and given way to new, revolutionary ideas. Not all of them wind up being correct (or even in the realm of possibility), but no one should ever be able to argue that physics is boring, Perhaps one of the singularly most interesting, not to mention complex, concepts to come out of theoretical physics is string theory.

To understand string theory, there are several things we must explore first, beginning with quantum physics. In traditional physics, we have so-called point particles and extended particles. The Standard Model of Particle Physics, which is the definitive starting point to all things subatomic, tells us that every particle known to exist is an extended particle except for three: quarks, bosons, and leptons. One exception is the electron.

Extended particles don't have a well-defined surface, instead, they are more like the atmosphere of the Earth, which is the thickest near the surface of the Earth and it gets thinner with altitude. Extended particles have a size, but the boundary or exactly where an extended particle ends is fuzzy.

Fermi Lab ponders, "Point particles are much more bizarre and are sometimes said to have zero size. This statement has raised more than one eyebrow. How can something have no size at all? And if it has mass, does the zero size mean it has infinite density?" The answer to that is no, for the record, but that's just one of the many mysteries surrounding point particles; They do, however, sometimes exist in quantum states that give them mass and charge.

It wasn't until we started slamming particles together at extremely high speeds that we began finding and classifying all of these different types of particles - the Higgs Boson should come to mind. We found that these elementary particles can be broken down into their further constituent parts, like quarks. Some, like baryons, exhibited characteristics of mass and spin (a baryon is a member of the quark and fermion family, which means it participates in one of the four forces: the strong interaction - something scientists were struggling to understand at the time).

However, the finding that really kick-started research into string theory was the discovery of a strange particle called the hadron.

As per the Cern Courier, "In the mid-1960s we theorists were stuck in trying to understand the strong interaction. We had an example of a relativistic quantum theory that worked: QED, the theory of interacting electrons and photons, but it looked hopeless to copy that framework for the strong interactions. One reason was the strength of the strong coupling compared to the electromagnetic one. But even more disturbing was that there were so many (and ever-growing in numbers) different species of hadrons that we felt at a loss with field theory – how could we cope with so many different states in a QED-like framework? We now know how to do it and the solution is called quantum chromodynamics (QCD)."

QCD, which was first developed in the '70s, is what pushed the concept of the strong force to emerge. It and quantum electrodynamics are closely linked. Both are considered to be examples of a special class of quantum field theories known as gauge theories - whereby particles theoretically interact as a result of the exchange of gauge bosons: "the photon for the electromagnetic force, W and Z bosons for the weak force, and gluons for the strong force." All, in turn, help comprise the all-important Standard Model of particle physics. Bringing us to...

The first version of string theory starts to come together:

String theory as a whole has many objectives: perhaps most importantly, it hopes to produce a "Theory of Everything," which includes figuring out how gravity works on a microscale. We've made some progress in that department, but here's a look at how it really got its start and how everything came together:

Gabriele Veneziano is the person who developed much of the framework for what would become string theory. He wrote an extremely important paper in 1968 that sent shockwaves through the theoretical physics community, It was called «Construction of a crossing-symmetric, Regge behaved amplitude for linearly-rising trajectories."

He worked for over a year, looking to develop some kind of theory to explain how the strong nuclear force (believed to bind protons and neutrons together, along with the constituent quarks and gluons inside protons and neutrons) interacted with hadrons. His work was largely mathematical in nature, but his calculations were enough to energize the community to better understand the implications of his original paper.

A few years later, three theoretical physicists - Yoichiro Nambu of the University of Chicago, Holger Nielsen of the Niels Bohr Institute, and Leonard Susskind, of Stanford University - made a huge discovery. Veneziano's mathematical proofs and general work trying to marry the strong force with the other forces could work if point particles were changed to string-like objects, which could contract, stretch and even vibrate.

String theory replaces all matter and force particles with vibrating strings that twist and turn in complicated ways and which look like particles from our perspective. For example, a string can fold and vibrate in such as way as to gain the properties of a photon, while another string folded and vibrating a different way has the properties of a quark, etc.

The problem with this theory was that it would require 26 different dimensions and it requires a particle to exist with no mass that travels faster than the speed of light - called a tachyon (which remains a theoretical particle to this day) to make his work consistent with relativity and quantum theory.

The three scientists mentioned above were among the first to suggest that changing particles to small loops would solve some of the problems with Veneziano's model. So what are these small strings made of, you may ask? Well, the answer is nothing really - the strings themselves make up everything.

One lingering problem was that the original version of string theory lacked an explanation for fermions, which have half-integer spin. That is until Pierre Raymond entered the picture and helped bring forth string theory 2.0.

Raymond reformulated string theory to include particle spin, allowing bosons, fermions and other particles with integral spins to join the club. Additionally, two other physicists, John Schwarz and Andre Neveu - from Princeton University - changed the game further by creating a version of string theory that didn't need to be consistent with quantum theory or special relativity. It also removed the need for the tachyon and 26 dimensions from the picture, dwindling the number of needed dimensions down to 10. This became known as the superstring theory.

A look at superstring theory:

Superstring theory (or sometimes called supersymmetric string theory) adds a few extra parts to the original theory, like doubling the number of particles in the standard model and replacing a few other parts. For instance, it says that all known particles can basically be divided into two types: fermions and bosons. The catch is, these two particles are very dependent on each other - there must be a boson for every fermion and a fermion for every boson. To add to it, every known and still-unknown-to-us particle must have its own "superpartner."

Symmetry magazine notes, "In the early 1980s, theorists realized that the Standard Model itself could be made supersymmetric and that this extension would resolve some vexing problems with the theory. For example, the small mass of the Higgs boson is notoriously difficult to explain—its calculation requires subtracting two very large numbers that just happen to be slightly different from each other."

Elodie Resseguie, a postdoc at the US Department of Energy's Lawrence Berkeley National Laboratory adds, "But if you add supersymmetry, this takes care of all these calculations such that you can get a light Higgs mass without needing to have such luck."

The biggest implication is that you can replace zero-dimensional elementary objects with 1-dimensional elementary constituents, or strings. The universe from the vantage point of superstring theory basically tells us that everything is fundamentally composed of these tiny strings. We perceive these strings to be particles based on how they vibrate when they are looped together, sorta like playing different notes on a violin or guitar.

It's important to note here that not all of the strings are looped together. Some of them are open (with two ends), while two other strings become twisted together, forming closed strings. A bit of tension must exist to help give rise to different modes of vibration, thus we can interpret these vibrations as elementary particles.

Open string vs closed string

Remember how I said that the mathematics for string theory originally only work with 26 dimensions of space time, but that was inevitably dropped to just 10 with superstring theory? That still sounds bonkers to people who appear to live in a reality with 3 spatial dimensions and one temporal dimension, but hopefully, we can make those extra details easy to understand.

History of extra dimensions and an understanding;

Two of the first people to suggest there may be extra dimensions - the originators of what would become known as the Kaluza-Klein theory - also happen to have helped lay the groundwork for string theory itself when they began working to unify electromagnetism with gravitation. For their work, they suggested that there exists not four, but five dimensions of space, However one of them is drastically different from the others, In fact, it might be curled in on itself through a process called compactification. It further implies that this compacted dimension only has gravity, and no electromagnetism,

David Berman, a Reader in Theoretical Physics at Queen Mary, University of London, puts it into perspective: «We are used to living in four dimensions. If you are arranging a meeting then you need to give one more piece of information: the time of the meeting, say 3 pm. With four coordinates you can describe any event. We don›t tend to clump time with the spatial dimensions but if you think about it, any event really happens in a spacetime with four dimensions. You can measure differences in time just like you can measure differences in space – we measure differences in space with a ruler and we measure differences in time with a clock. So anything you can think of in terms of space you can think of in terms of time.»

"We can learn a lesson from this: our human perception of dimensions is limited. We only perceive three dimensions. We can understand that time is an extra dimension, an extra location for any event, but we don't see time in the same way. We experience it very differently from spatial dimensions as human beings. This hints that human perceptions of things may not be the end of the story in terms of what's possible."

"Mathematically the number of dimensions is just the number of coordinates you need to specify a point. We're familiar with specifying points in four-dimensional space time, but can you imagine a space where you need five bits of information? Or six? Or seven…? Mathematicians regularly work in higher dimensional spaces, but they are not the only ones. For example if you're a sound mixer making music you might be working with 12, 24, or even 128 tracks. At each moment in the music each of the, say, 24 tracks has a specific volume, defining a point in a 24-dimensional space of sound. The number of dimensions is just the number of bits of information you need to specify a point."

Confused? Yeah. Kaluza and Klein's research didn't immediately become the subject of speculation - its explanation for quantum gravity was weak. Moreover, as quantum physics further developed, it didn't account for the strong and weak nuclear interactions. However, new life was breathed into it when the Yang-Mills theory was developed in the 1970s. This was an attempt to unify all the fundamental forces except gravitation - ultimately unifying the weak nuclear interactions with electromagnetism. It can be thought of as the most symmetric field theory that does not involve gravity.

Other theories then started taking shape... Like,

String theory on steroids: Superstring theory

Superstring Theory, or supersymmetric string theory, is what you get when string theory and the concept of supersymmetry combine. If you're not familiar with supersymmetry, gather round..

In the standard model of particle physics, there are essentially two groups of fundamental particles: We call them bosons and fermions. The former mediate the fundamental forces as integer spin particles. while the latter have half-integer spin and comprise matter.

Most theoretical physicists believed the bosons and fermions to be connected in some way, but the math suggested otherwise. This is why the idea of supersymmetry was born.

As a primer, check out the video below, where an expert from Fermilab helps explain how the concept of supersymmetry works in the most simplistic way possible:

According to Cern, "Particles like those in the Standard Model are classified as fermions or bosons based on a property known as spin. Fermions all have half of a unit of spin, while the bosons have 0, 1 or 2 units of spin. Supersymmetry predicts that each of the particles in the Standard Model has a partner with a spin that differs by half of a unit. So bosons are accompanied by fermions and vice versa. Linked to their differences in spin are differences in their collective properties. Fermions are very standoffish; every one must be in a different state. On the other hand, bosons are very clannish; they prefer to be in the same state. Fermions and bosons seem as different as could be, yet supersymmetry brings the two types together."

It does this with a new particle, the Higgs boson. While the Standard Model seems to predict that all particles should be massless, this is at odds with experimental observations. The Higgs boson provides this mass. At the same time, "the extra particles predicted by supersymmetry would cancel out the contributions to the Higgs mass from their Standard-Model partners... The new particles would interact through the same forces as Standard-Model particles, but they would have different masses. If supersymmetric particles were included in the Standard Model, the interactions of its three forces – electromagnetism and the strong and weak nuclear forces – could have the exact same strength at very high energies, as in the early universe."

The new theory also predicted most of the known particles and a new spin-2 particle, the 'graviton', which is a candidate for gravitational force-carrier.

Over time, physicists came up with five different versions of Superstring Theory, namely Type I, Type IIA, Type IIB, Heterotic, and Heterotic with E(8) x E(8) gauge symmetry.

The first superstring revolution attracted a lot of attention. In 1995, Edward Witten, a theorist at the Institute for Advanced Study in Princeton, New Jersey, argued that the five string theories actually each represented an application of a more fundamental, 11-dimensional theory. This unified approach is popularly called 'M-theory'.

Extra-dimensions of string theory

We left our discussion on extra-dimensions incomplete while talking about Superstring Theory. Let's continue.

String Theory supports 10 dimensions, 3 of which extend indefinitely and are observable to us. So, where are the other 6 spatial dimensions?

They are right here but curled upon themselves, i.e. they are compactified.

The extra-dimensions are inherent in the mathematical notion of 'manifold'. Take, for example, a relatively large sphere, and place an ant over it.

The surface of the sphere will look flat to the ant. So, we have a 'local' shape (flat surface) and a 'global' shape (the sphere) here. The sphere, in this case, is an example of a 2-dimensional manifold.

It is the same with our world.

We live in a 3-dimensional manifold. Even if we do not consider the postulates of String Theory, general relativity suggests a fourth dimension, gravity, and the universal phenomenon is described with the help of curvature of this additional dimension.

Thus, we can easily conclude that a manifold may have curvature and other non-trivial properties.

A Calabi-Yau manifold is a class of 6-dimensional manifold and is a subject of study in String Theory. They wonderfully predict several realistic theories in 4-dimensional space-time.

We have drawn heavily from the earlier discussed Kaluza-Klein idea of compactification. The extra dimensions are just compact manifold which is too small to be detected by us.

Furthermore, M-theory extends on string theory's 10 dimensions and works with a total of 11 dimensions.

In Closing: Is string theory really a "theory of everything?"

The Standard Model of Particle Physics, while a very important part of understanding how the universe works on a microscale, still isn't completely without flaws. Yet, it's definitely our greatest tool when it comes to putting together a decisive model of string theory.

Most importantly, we have quantum mechanics, general relativity, quantum field theory and many more models to gain insight on the same subject that is the Universe.

Can't they all be connected in some way? Shouldn't quantum mechanics and gravitation come under a common framework? Yes, they should.

The Theory of Everything is our attempt to unify different theoretical models to probe Nature. String Theory came out as the strongest candidate for such a unified approach.

It has unexpectedly predicted the quantization of gravity with gravitons. So, why isn't it the ultimate discovery?

String Theory has been somewhat successful in explaining many complex phenomena, most importantly black holes. Black holes are very small objects with very large mass and it requires general relativity as well as quantum states to study them.

String Theory has also offered new insights into quark-gluon plasma and a relationship between quantum field theory and string theory has been established, called AdS/CFT correspondence.

String Theory has produced numerous results, some of which may seem absurd or incomprehensible. For example, it has been used to predict the existence of 10^{500} universes or a massive multiverse.

However, String Theory has faced many setbacks. String Theory models rely on supersymmetry which, like the extra dimensions, has yet to be observed. The models are also criticized for their inability to adequately describe an expanding universe.

But what startles the scientific community is the recurring comeback of this controversial theory. Perhaps largely to do with the usefulness of the mathematics involved, which has highlighted connections between different areas of math, inspiring many new ideas and approaches.

String Theory has been the object of interest in the media and popular science as much as in the scientific community for its counter-intuitive solutions. The biggest problem with String Theory, however, is that most of its predictions and notions can't be tested with experiments, as they require very high energy, which is currently not possible with the tools we have.

But String Theory can also be thought of as a revolution in the world of physics, and it is likely to be a part of mainstream physics in one way or another.

So, here is String Theory simplified.

In string theory, the *multiverse* is a theory in which our universe is not the only one; many universes exist parallel to each other. These distinct universes within the multiverse theory are called *parallel universes*. A variety of different theories lend themselves to a multiverse viewpoint.

In some theories, there are copies of you sitting right here right now reading this in other universes and other copies of you that are doing other things in other universes.

Other theories contain parallel universes that are so radically different from our own that they follow entirely different fundamental laws of physics (or at least the same laws manifest in fundamentally different ways), likely collapsing or expanding so quickly that life never develops.

Not all physicists really believe that these universes exist. Even fewer believe that it would ever be possible to contact these parallel universes, likely not even in the entire span of our universe's history. Others believe the quantum physics adage that if it's possible, it's bound to happen somewhere and sometime, meaning it may be inevitable that quantum effects allow contact between parallel universes.

The idea of a physical multiverse came later to physics than it did to religion and philosophy. The Hindu religion has ancient concepts that are similar. The term itself was, apparently, first applied by a psychologist, rather than a physicist.

Concepts of a multiverse are evident in the cyclical infinite worlds of ancient Hindu cosmology. In this viewpoint, our world is one of an infinite number of distinct worlds, each governed by its own gods on their own cycles of creation and destruction.

The word *multiverse* was originated by American psychologist William James in 1895 (the word "moral" is excluded from some citations of this passage):

"Visible nature is all plasticity and indifference, a [moral] multiverse, as one might call it, and not a [moral] universe."

The phrase rose in prominence throughout the 20th century, when it was used regularly in science fiction and fantasy, notably in the work of author Michael Moorcock (though some sources attribute the word to the earlier work of author and philosopher John Cowper Powys in the 1950s). It is now a common phrase within these genres.

According to MIT cosmologist Max Tegmark, there are four levels of parallel universes:

- **Level 1:** An infinite universe that, by the laws of probability, must contain another copy of Earth somewhere

- **Level 2:** Other distant regions of space with different physical parameters, but the same basic laws

- **Level 3:** Other universes where each possibility that can exist does exist, as described by the many worlds interpretation (MWI) of quantum physics

- **Level 4:** Entirely distinct universes that may not even be connected to ours in any meaningful way and very likely have entirely different fundamental physical laws.

So for I have discussed, I need now only one concept to end up my research paper and that is the concept of multiverse and yes, it is the backbone of my paper and it is the same thing that is responsible for the solid conclusion of my thesis but you need to wait a bit till I explain to you some basic things. I hope you would feel it the way I feel it and wait till my resource paper gets published

Now question arises in what respect does the natural selection fall in the relation of blindness, or how does the blindness of natural selection arise? from the very beginning of this universe till its ultimate end, all these rules and laws hold true whether we talk of very big bang or talk of the existence of solar system or anything else or beyond that we see that these rules and laws are adhered and obeyed by and stand the ultimate reality and driving force the lie behind every natural phenomenon.

since the famous big bang has taken place, the universe is continuously expanding. It is too followed a definite path that are the laws and rules and it would go on following the same laws and rules till it's last stage. From very beginning up to its end it goes on following the same laws, rules and. principles and eventually, it final stage will be shaped on account of these rules and laws it abides by and in case, it would have not been consistent and systematic it would definite have taken a short out reach to its final stages without caring a bit of the laws and rules and principles. would not it have? after all, question now, ultimately evolves and strikes why would it follow the laws and rules? it would tend to came to its ultimate end if its future, that is ultimate end is pre-decided or presumed, rather, eventually, shaped, as we know, then why won't it breach any of the rules or principals? what sort of insanity it is? So, it re-affirms and confirms, and appropriately, endorses our view that these rules are just blind knowing hardly about their being, about their existence and yes about their future. this is because of such blind nature with respect to its rules and laws that it still goes on following the rules and laws to reach its goal, ultimate end, systematically and in a pre-designed fashion. This is because of these two rogues rules and laws, that natural selection and, yes, ever expansion of our universe continues to be. These rules and laws compel it to continue.

These are just the rules and laws that fashion, that decide what to do the next or what would be the next or the final course of anything. So rules, laws and their outcome I.e natural selection are blind but as I have consistently claimed and admitted with extra ultra high-tech senses. This can be done but with extra high-tech sense now the question arise what are these extra high tech senses.

Time properties

Mass texture Behaviour

Energy structure

Space

Energy mechanism has its own outcome, properties and behaviour. This all decides the mechanism, mass energy and properties. And behaviour of these fundamental units is all decided by the mechanism.

Let's Take an Example of Heisenberg Uncertainty Principal

The Heisenberg's uncertainty principal is one of a hand full of ideas from quantum physics to expand into general pop-culture. It says that you can never simultaneously know the exact position and exact speed of an object. Uncertainty is often explained, as result of measurement that the act of measuring an object's position changes its speed, or vice-versa the real origin is much deeper and amazing than the uncertainty principal exists.

$$\Delta p. \Delta x > p/4\pi$$

Because everything in the universe behaves like both a particular and wave like at the same time. In quantum mechanism, the exact position and exact speed of an object have no meaning to understand this. We need to think about what it means to behave like particles or a wave particles by definition, we can represent this by a graph showing possibility of finding the object at a particular place, which looks like a spike 100% on a specific position and zero everywhere else. waves, on the other hand are disturbances spread out in space.

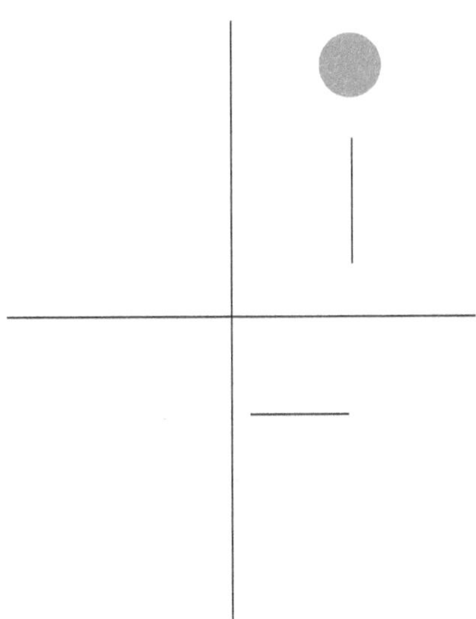

Like ripples covering the surface of a pond.

We can clearly identify features of the wave pattern as whole, Most importantly, its wave length

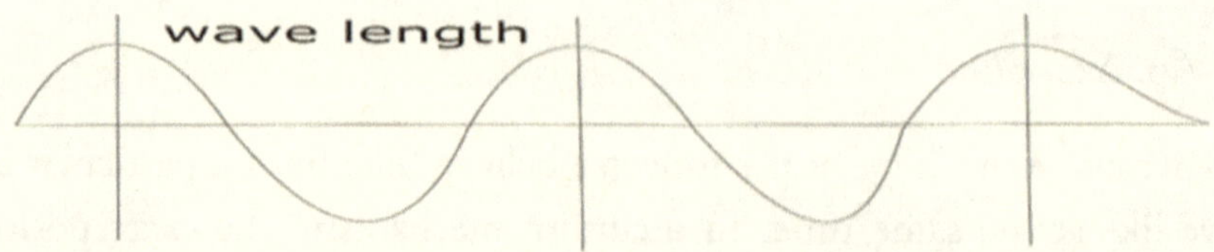

which is the distance between two neighbouring peaks or two neighbouring valleys, but we cannot assign it a single position

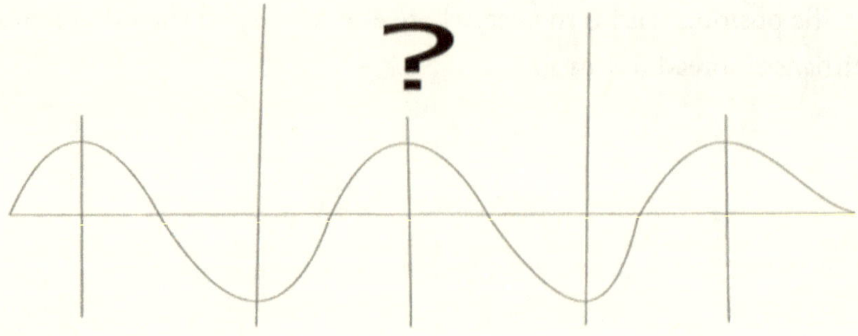

but we cannot assign it a single position it has a good probability of being in lots of different places. Wave length is essential for quantum physics

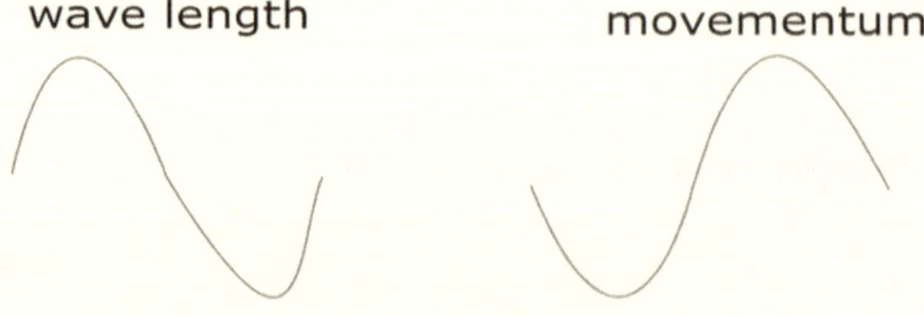

Because an objects wave length is related to its momentum

M × V

A fast moving object has lot of momentum

Which corresponds to a very short wave length, a heavy object has lot of momentum even of it is not moving very fast which again means a very short wave length. Small things like atoms or electrons, though, can have wave lengths big enough to measure in physics experiments. If we have a pure wave we can measure its wave length and thus its momentum but it has no position we can know particles position very well but it does not have a wave length so we don't know its momentum to get a particle with both position and momentum to make a graph that has wave but only in small area.

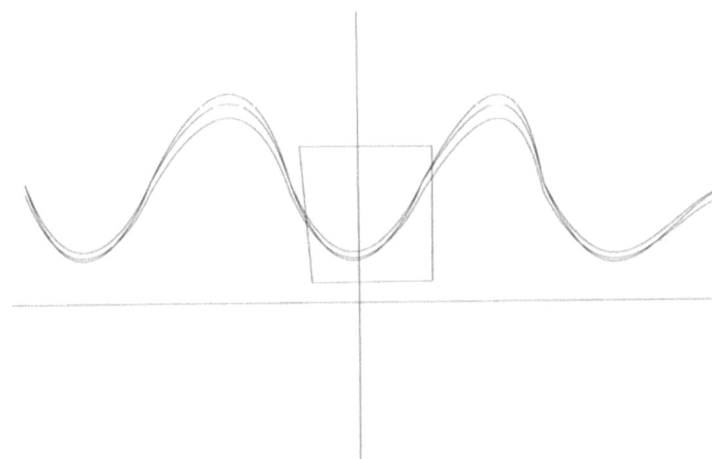

How can we do this by combing waves with different wave lengths,

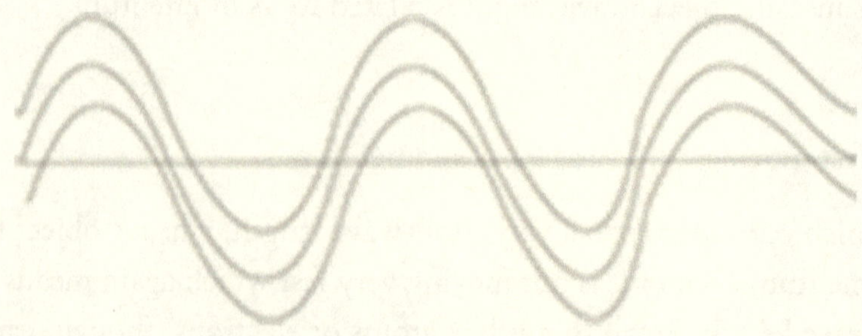

which means giving our quantum some possibility of having different movements when we add two waves we find that there are places.

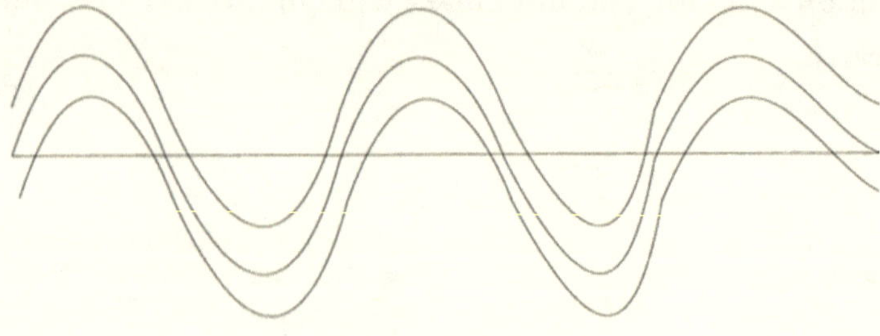

Where the peaks line up making, a bigger

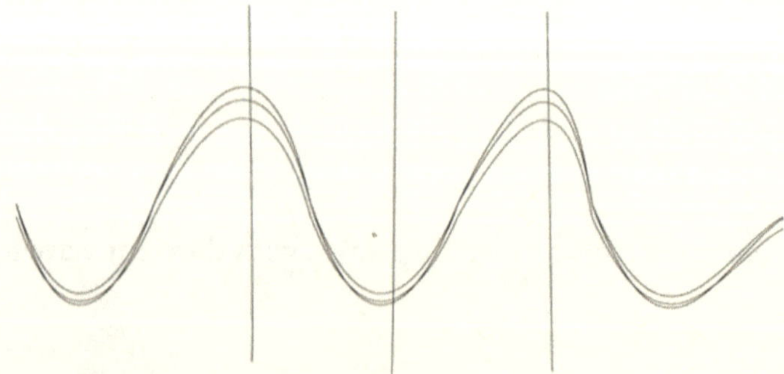

wave and other places. Where the peaks fill in the valleys, the other results has regions where we see waves.

Separated by regions of nothing to all. If we add the third wave the regions where the wave cancel out get bigger.

If we keep adding waves we can make a wave packet.

With a clear wave length, is one small. Region that is a quantum object with both wave and particle nature. But to accomplish this we had to lose certainty about both the positions and momentum the position is not restricted to single point. There is a good possibility of finding it within some range of the centre of the wave packet and we made a wave packet by adding lots of waves which means there is some probability of finding it with the momentum corresponding to any one of those both position and momentum are now uncertain and uncertainties are connected. If you want to reduce the position uncertainty by making a small wave packet you need to add more wave which means bigger momentum uncertainty. If you want to know the momentum better, you need a bigger wave packet which, means a bigger position uncertainty this whole represents the behaviour of moving particles.

After all what is this behaviour.

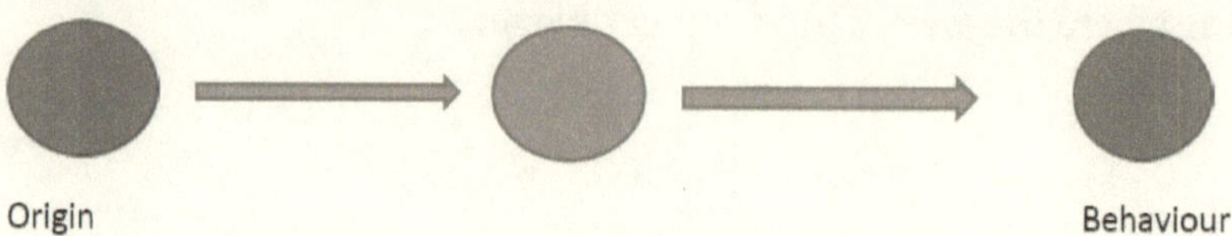

behaviour is just the outcome of various steps- if we succeed in vanishing these steps, is there any possibility to reach to the existence of nothingness?

the answer to this question is very dynamic and is, 'yes'. we can reach to the nothingness then, and that nothingness would exist in the reserve state. We call that reserve state- nothingness, as idea.

And one more dimension, the same thing makes up the sense. Now, there various /many steps function as per the sense and thus they go on deciding what the next step would be like. The whole universe is filled up with a quantity that we call as information. What else our observation is or what else we observe that is all subjected to the '3d' perspective and who knows what sort of behaviour would be there for the same quantities in 4d, 5d, 6d etc?

(12) Now the very logical question gets its genesis and that is if the behaviour get opposed /changed in different dimensions?

Let's take an example of Free Energy! why after all is this impossible for us? The word 'amount' has a greater bearing on Free Energy?

you already know, kinetic energy is given by

$K.E = 1/2mv^2$

The absence of Time Factor in this equation means a lot in the disformation of Free Energy. And if we exclude all the forces that can be exerted on a body- so consider an empty space and a body 'a'- which is a bar magnet. The body will always tend to be in motion (if it's originally moving)- This keeping of its to be in motion always doesn't mean it's having unlimited energy. NO- It does have a definite amount of energy- and if we bring a coil in the vicinity of that body, That means we are taking energy out of it and thus decreasing its speed.

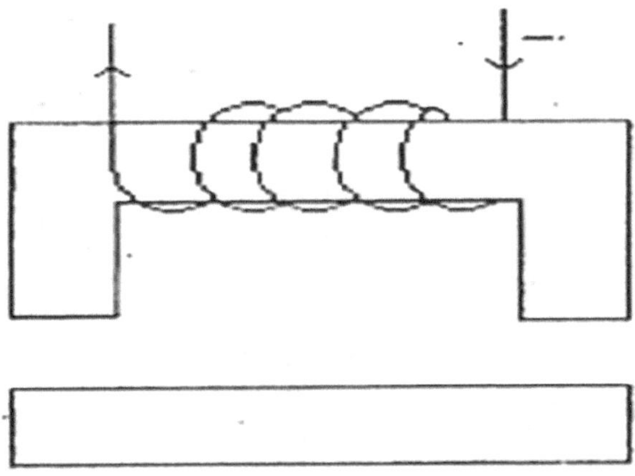

but note that this process is demonstrated by us in '3d' perspective.

Would it show the same behaviour if we move to 4D, 5D, 6D etc.?

You know electrons go on revolving around the +vely charged nucleus in the atoms- they do have kinetic energies. But the total energy in an atom is always constant- that does have specific value (Amount) please be away from the misconception of any sort that motion (continuous motion) points out to an unlimited energy- This is only an illusion- that we have been compelled to observe on the part of the universe- this is the thing that strikes on common sense. The continuous motion is there just to prevent physics or physical laws from deception.

what is the significance of life time motion with unlimited energy- It's only that much as to conserve energy and stabilize physics. This is only a tool of physics. so, in nutshell, what I want to convey to you is that: time-space framework provides an object a specific meaning- otherwise, there is nothing that gets added or subtracted to the object with respect to the information that exists with it.

let's take another example of free energy-

you know of the atom.

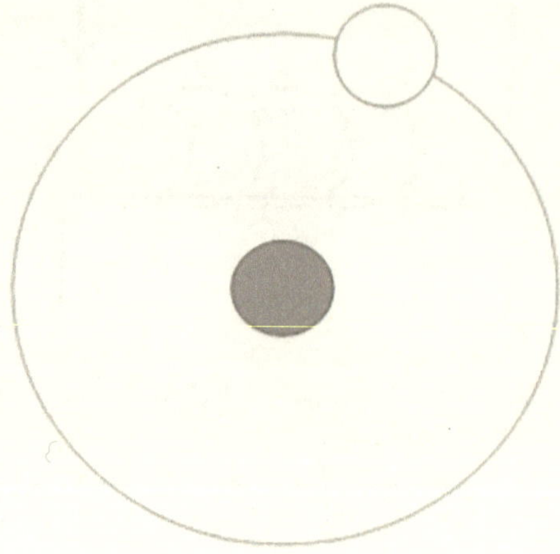

The electron moves and moves around- if we replace electron and proton with magnets- 1st be confident that a magnet can move around a magnet- but we can't derive any energy out of it- This is because motion is only an illusion

On Ideas

Idea is a foundation but it is blind

Idea creates a sense of presence. Presence Possess, Texture.

Texture possess behaviour.

Behaviour possess properties.

➤ Properties create functions.

➤ functions take shape of laws and rules.

The first reason or secret behind the creation of this subtle universe was the possibility of its coming into existence and it, too, was necessary that what was and what is going to happen to this universe in its ultimate end, the existence of its possibility and similarly, the possibility of its end must exist to that is what is going to happen.

Edison has once said, " I have not failed I have just found 10,000 ways that won't work" now the question in vogue is and should be, whether this universe goes on as per its aim /ambition or logic or does it simply goes on following the natural selection process? I have just a-head quoted Edison. Now the ultimate question will strike that is, has the universe experienced all the logical possibilities that it would not have seen the present universe?

Electric Current

Let us take example of electricity

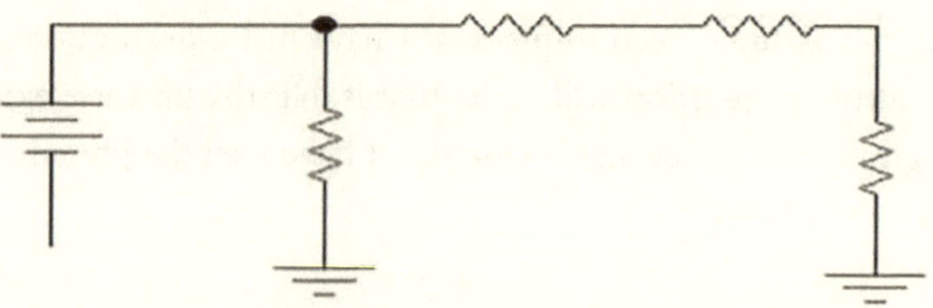

We know of the electric current that electricity moves from its high potential to the low potential in a circuit with almost no resistance as

v = IR

when resistance is high it is difficult to move the

v = IR

V/R = I

potentials, so current takes the path with minimal resistance the above diagram depicts all. Now the question arises; does current 1st go through a medium with resistance or does it directly go through the circuit?

How is this possible?

Electricity knows the medium and goes through the path with minimum resistance when the switch is on.

The current flows without potential difference through the circuit. Does this mean or does this point out that we can discuss and define the current without potential difference? And, it is the same potential difference that is the determining factor of the existence of current. And to observe current, potential difference must exist. Now we will look into the

Case 1st

When current flows through the circuit that we have designed for it their are two ways, we have provided for the current to flow through.

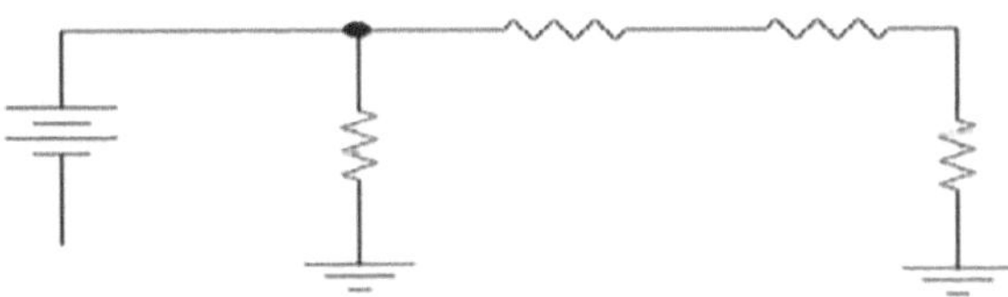

Now which path the current will take to flow this to the question that arises.

we have observed that most of the current flows through the path and the path second is followed by the least current

the 1st question got solved here that is maximum current will follow the first path and the minimum current, the second path. Now another question strikes the mind and it should logically exists as well, that is how? Now there can be two solutions or answers for it.

1st answer

either this current follow's the minimum resistance path already.

a second answer

the rules of current determining the path should take to flow.

Case 2nd

and, or the current ist tries every path to follow and then flows the path with minimal resistance value.

Case 3rd

or is it because of the rules and laws that determine the behaviour of the current.

And why all this? This is what sense of physics is all about. Now let's try to determine the sense.

the rules and laws of physics are the pre requisite for the sense. And sense is the outcome and let's try to unfold the mechanism

Path of current as discussed, so many times, this current followed the path with minimal resistance and the current follows that path that's connected to the ground.

Most of the current follows the ist path that has the least resistance amongst these

Now let's create a situation

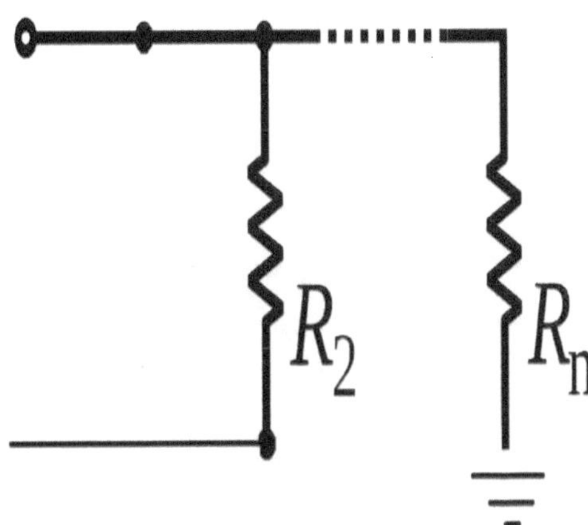

Here the current will take the path that is connected to the ground. Now, we got one more information from it. After analysing these images/ experiences we come to the conclusion that the flow of the current tends to be towards the ground, the purpose of the current is not to do any work but just an err, on its part, compels it to do that. And it is the high and low potential difference becomes an opportunity for us to do some work from it.

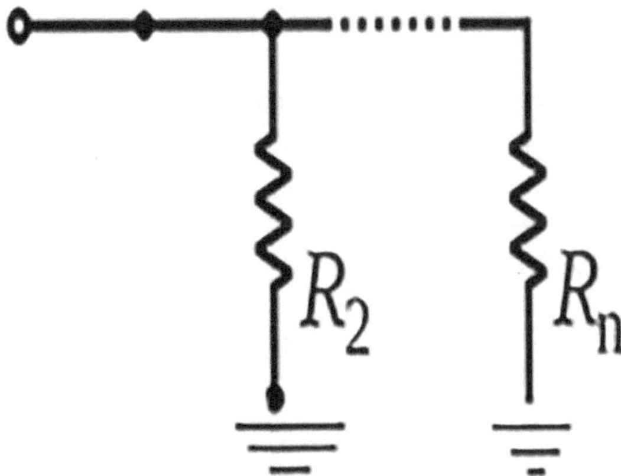

Now try to understand this, we have put the load along its path and it becomes its compulsion to get our work done

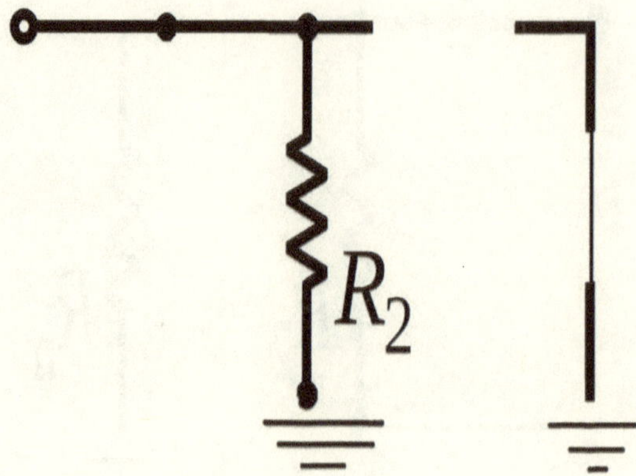

this is sense but it is just the compulsion, I may tell you that physics is hungry of the results. Every process wishes a results

And this result becomes one more reason for the work done, its compensation is our gain. I think, it's injustice with it on our part in a sense —bad joke anyways.

When we fulfil all the requirements, then what will the results of a process?

What demands a process? could/should we give that result a name such as stability?

Yes, that is called the stability with respect to the demand of the process, a condition prevails, that we call as stability.

Behaviour vs Sense

sense is a property; behaviour is the outcome. sense determines the behaviour of physics. Now the question arises, when does the behaviour precede the sense? Let's try to talk of the case first: - when we switched on, the current was distributed equally through both the paths that is the path first and path second and when the resistance came to foreplay its role, then the major amount of current tried the way/path with minimal resistance value. And in case, the sense persists, it would not have, at 1st, followed the paths with equal distribution of the current. Rather, greater amount of current would have passed through the minimal resistance path and lesser amount of current would have followed the path with maximum resistance. So, the result as we have mentioned that it is laws and rules that determine the sense of physics. The sense of physics has never been blind when no inventions were there. when we tried the new things or new inventions to come into being, the laws and rules of the physics went blind otherwise, they were not blind rather a well-developed mechanism was on its natural flow that too, from the very existence of the universe.

Now types of senses in physics

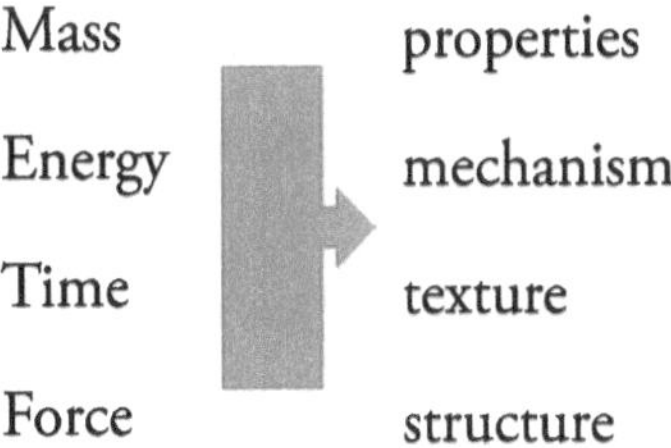

The properties of the above mentioned fundamental quantities, their behaviour and their very inner mechanism, texture and structure have given the genesis to the present day universe that is why we have not been able to be-fool the laws of thermodynamics with due our invention. we give existence to our invention with due existence of this universe and so we are a link between the two. we run side by side and it is, still that we are tied together with respect to each other.

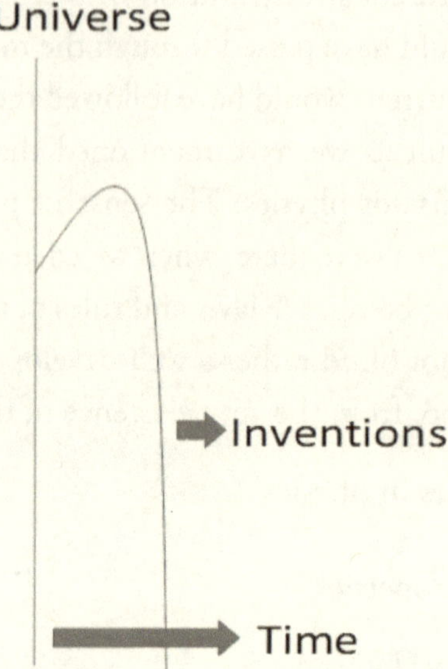

The inventions are possible only on account of the existence of this universe and that is why it is an uphill task, rather, impossible to conquer the laws of the thermodynamics. Because if we could do it, a new existence of a new universe begins.. A new 2nd universe will come into being

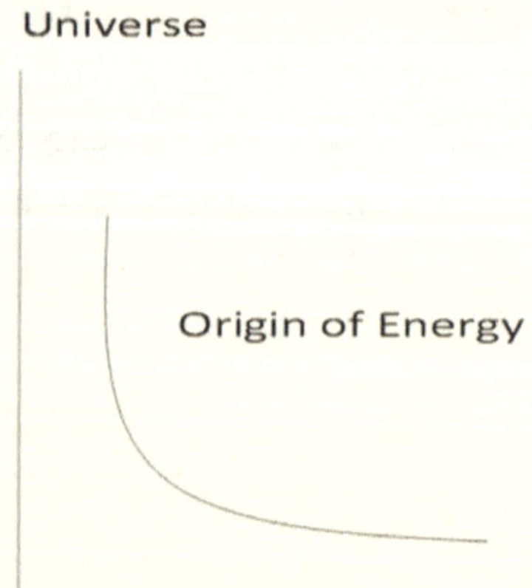

Inventions by Nature

Nuclear fusion

Stars are powered by nuclear fusion in their cores mostly coverting hydrogen into helium, deuterium and tritium isotopes of hydrogen.

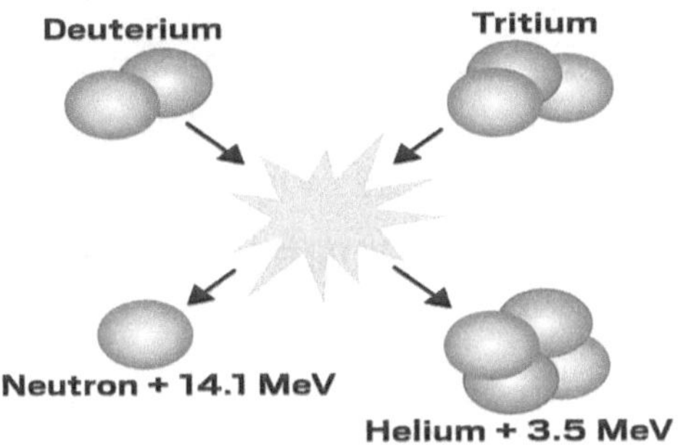

This process is an invention by nature.

First we'd know the things necessary for inventions.

Possibility and Impossibility

If the things would not have been impossible then the definition of possible would have been utterly different. the name should have been magic then. As there are no laws no rules to magic. Because rules and laws make up the things. So, they make up the words, it's quite too observable a phenomenon. The inventions create inventions. We human beings are the inventions of nature and we create inventions.

Importance of possibility in universe

Impossible makes things possible. Impossible creates one single way up to the results that is the impossible creates a way to make things possible.

Impossible act by the potter's hands gives rise to the magnificent shape of the utensils.

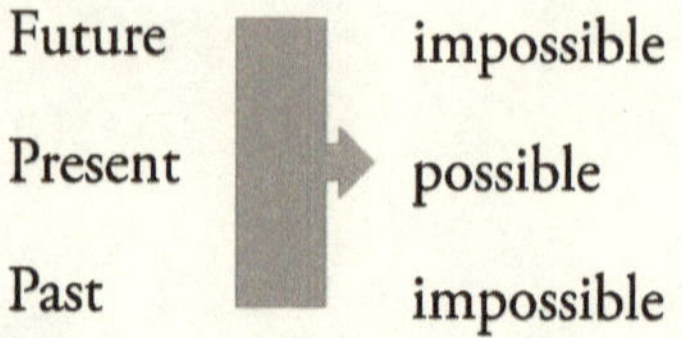

Impossible occupies all the unreal things. We can see a thing becoming possible that we earlier feel to be impossible. So we came up to the observation that the same thing becomes possible that we treat to be impossible otherwise. So, the question is; are we able to see or observe the impossible becoming possible and the answer is 'yes'! because the impossibility is the hindrance or the barrier and we can easily see the barriers and hindrances. These are the same things that converts the impossibility with respect to the thing into possibility.

Let's take an example of electromagnetic induction

I have set up a system, I have taken up the positive move of the coil to the magnetic field but the current did not generate because.

$$W = FSCos\Theta$$

When $\Theta = 90$

$$W = FScos90$$

$$W = 0$$

that is work done is zero. So, no current is generated for $\Theta = 90^0$. If we change the magnetic field, keeping the setup the same, that is changing the angle Θ, and the current is generated. How? That is it is the angle that decides whether the current generates or not. That is $\Theta = 90^0$. is the hindrance or the obstacle -this single hindrance, though occupied its single way (existence) but it gave way to the numerous other ways to get the work done. That is, this simple impossibility gave way to the possibility of many other things.

so one impossible way moves around infinite possible ways. And in case, their are innumerous impossibilities, Think of the possible ways how innumerous they can be? And that is the reason why the existence of the invention is too tough. Passing through thousands if not millions of impossibilities to get to the inventions. And these impossibilities are what the world or universe exists as it is because if there wouldn't have been any impossibilities - natural check mates, this universe would have been just a fantasy place to live in. Next important dimension of the inventions is-the time. As much the time taken by the inventions, So much subtle they are! and if it is done in hurry or jiffy, so much is the chance of the errors in its existence

the nature 1ˢᵗ gave genesis to the universe and then it was our turn to get created.

Universe human

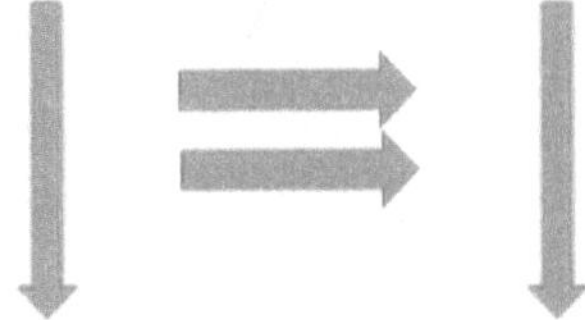

Invention invention

All our inventions are universe -based that is they fully depend on the universe. if universe perishes, our inventions be no more. so, it is the universe that is of utmost importance for the inventions on our part to take place.

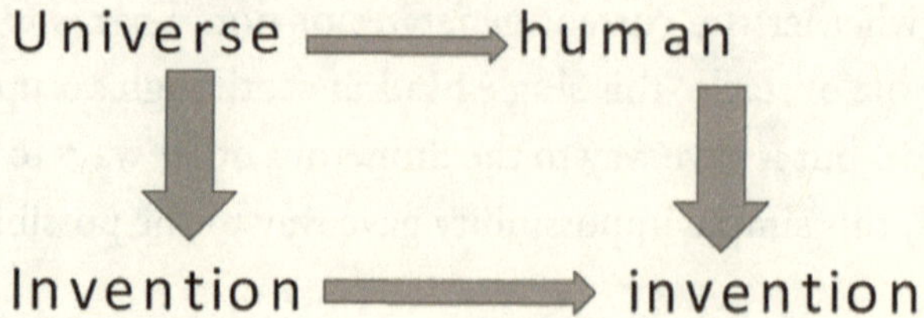

so the universe has the direct bearing or links with our inventions. it allows the same to happen what its programming is or as per its own senses, as per its own information and the possible is what can happen as per the information. Now the question shifts to the next level that is the information of the universe. let's try to pay a thought to it.

Information of the Universe

Here let's put our heads into the sub-atomic particles like photon, it is made up to of two down-quark and one up-quark with mass 1.10 those quark possess only 1% of mass of the photons whereas the remaining g mass is just the information. I may be wrong here but I believe it to be true.

Now let's come to the de Broglie's wave equation.

$$\lambda = \mathbf{h}\,/\mathbf{p}$$

And if we break equation into h,m,v

M ⟹ Mass

V ⟹ Velocity

h ⟹ planks constant

and if we increase the mass of the body, the wave nature decreases. we have just increased the mass of the moving body, and with this we have changed its behaviour. you can see, how just the mass- change resulted in the alteration produced in the whole behaviour of the particle. It's sense admitted to change whole of its behaviour. Rather, its sense prepared itself to change its behaviour a milli second earlier thus

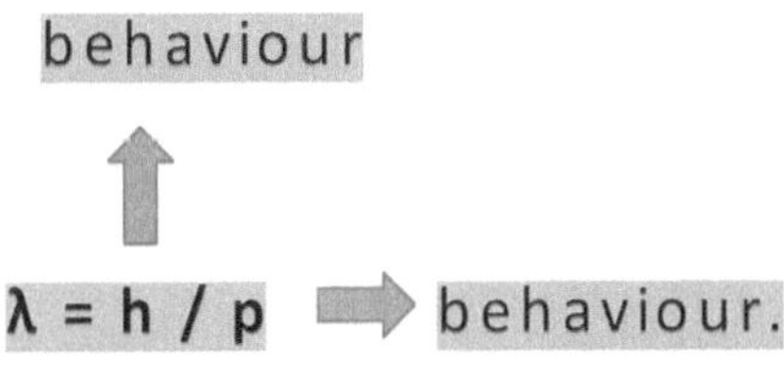

The product of mass and velocity here gives the genesis to its existing behaviour.

Behaviour + behaviour

⬇

Sense

⬇

Behaviour

Before the creation of the universe, behaviour could have been observed or the sense?

Yes, the existence of the sense could have happened. But just after its creation, going beyond or up the atomic level, we can perceive the behaviour and not the sense, and, in case the sense precedes the behaviour, it means the chance played its part. it's just the chance and nothing else.

And when the sense comes 1st, it means that there is something behind it, something of the information to happen the way it happens. It is the solid information, the rules and laws prove to be wrong only when they are derived from the behaviour 'now tell me truly if or whether you feel something'? Now let us, talk again of the origin of the universe we have talked of the quarks, we have talked about the mass of the photon, the mass present in the photon in the form of the information, that will be well regarded as the sense of physics. it means that ''sense'' too is the back bone of the present day universe.

Now let's deal with impossibility

The first and foremost feature of impossibility is that;

impossibility creates a straight path. and as you know the straight path, if it is between two definite points it becomes the displacement. Impossibility, actually provides the boundaries- two definite points, to form distance and displacement.

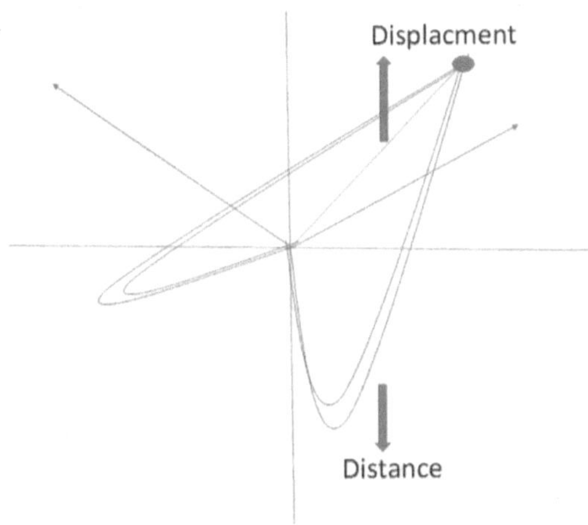

let's take coordination of points A and B.

It is quite evident that there are various ways/paths that can join A to B. But when we come to find the shortest of all these paths, it is the one and only the one. This alone is the 'possible' one. Hence it is the displacement. And, in case, we try to find other ways/paths from A to B without any regard to the distance or length, then there are various, infinitely many ways that are possible there to join A to B and these all do we regard as the distance and the distance follows every geometrical path that is, it is irrespective of the direction by parallelogram, triangle, square, rectangle, hexagonal, parabola and irregular ways as well. And the way quite opposite to the direction of the point will be straight. And the ways excluding the point remaining in other directions excluding the displacement, will be kept constant for future by nature. It connote that there is just a single way that falls in the realm of possible that is the one represented by the displacement.

And the rest of the ways/paths joining A to B non-linearly, are just impossible ones but they keep on changing to the possible ones as we move on.

illustrations: suppose we take a point A of three dimensional axis with x,y,z coordinates in the vicinity of other points B, C, D etc.

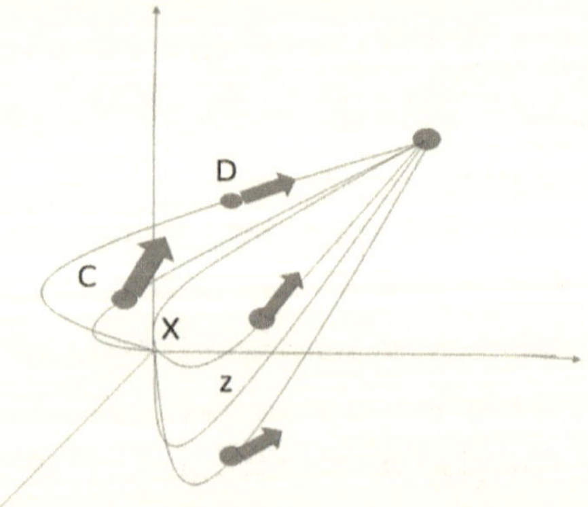

if you check closely, you come to know that every point shows a direction that is not possible that which was the motive- all that happens here is impossible. But if we go on analysing we get surprised on learning that impossible turning out to be possible by and by and that which was impossible, makes the things become possible.

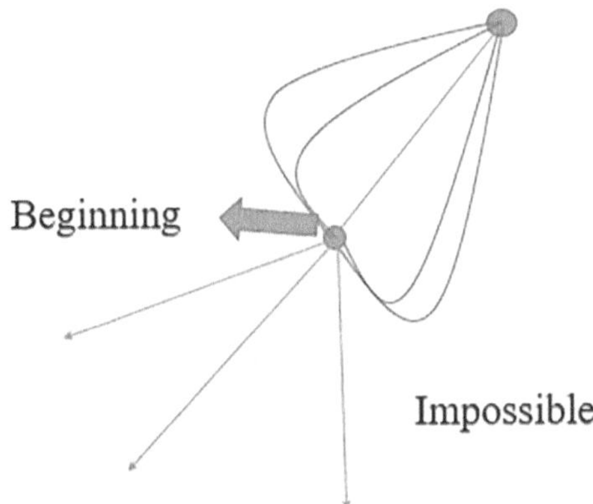

it is these impossible ways that keep their path occupied. and that which is possible does not need to follow the impossible one. It exactly acts like the potters hands, which gives shape to the mud and make a beautiful art. There is no result related to the impossible and this is what gives the possible a strong existence and quite adds to its meaning of being.

now let's illustrate it with an example you just consider yourself as in some other planet, revolving around its axis and this sun, like on the earth you do- but without any inventions! there are, otherwise, such inventions that were expected but some other things came into being.

yes, results can be different from the one and the same mechanism

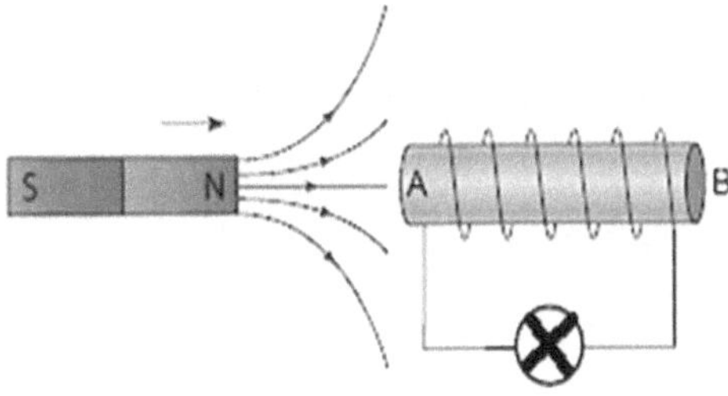

as per the Faraday's laws, the changing flux creates emf. We have taken a coil, here, and connected it to a bulb. we moved a bar magnet, and the current got generated so we created a situation with the installation of a bulb and that bulb lighted. This gave us a result in the form of lighting of the bulb. Now, let's create one more situation, replace the bulb with a speaker, when the magnet is in moved, the sound gets produced.

Hence, you can see the two different results for the same process

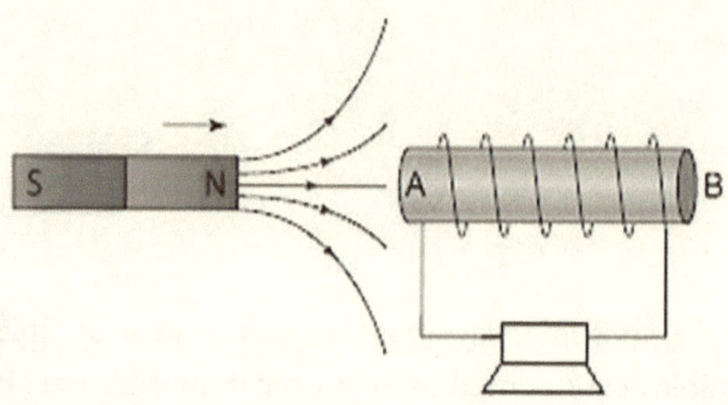

This,-bearing different results from the same process is the hidden principle of laws of thermodynamics. That is, we can convert the energy.

The topic here is that external factors do have a great relevance on the results.

The main objective of creating/making the universe

Now we do know the starting we do have observed it but what is its ultimate going to be- is the thing that we don't know yet- the actual and the most important question is- that the path that is followed by it- (The universe-) is whether the distance or displacement- that is impossible or possible. Now let's suppose or let's imagine its end going to be as in the figure..

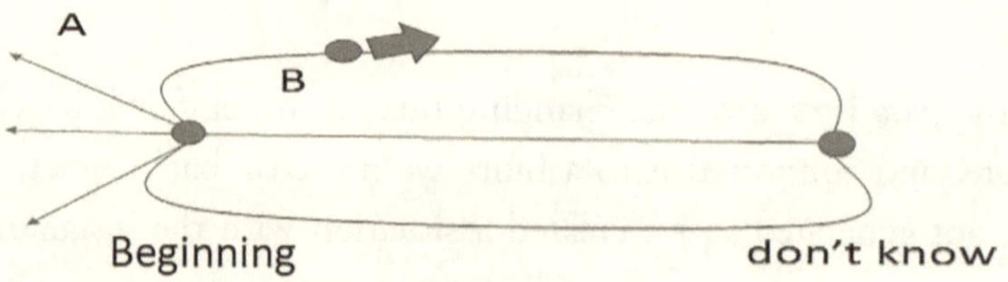

Now the question arises whether it follows the path A or path B or path C. And that which is happening and that, too, which has happened, is that a distance or displacement whether to do was something else and that which is going is some other thing- is this universe following the same path that is pre-programmed into it or not? This together with all other important questions, it is the endeavour of physics to get into such tricky and methodological questions

Rules and laws to determine what the path should be

This is, what we call as, the sense of science let's analyse the figure once more

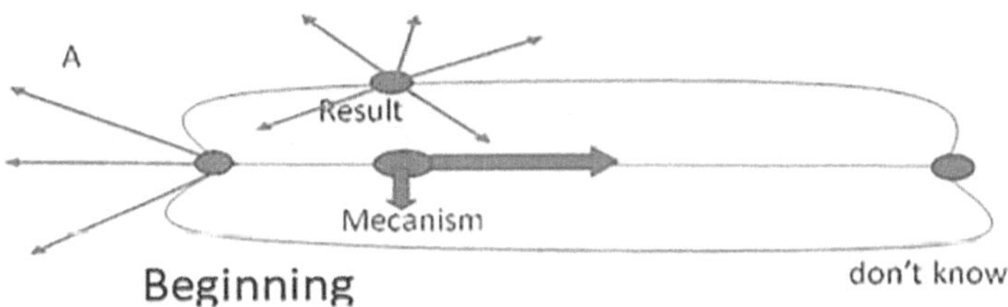

Here point A determines, what is impossible and the step after it denotes the sense of science. And that is what we call the laws and rules. Mechanism tends to give results and that is always directed to the point - do not know. And all other directions except this point are in a constant wait for the results. And, in case, the results come out that too, in two different ways then? It is the shortcut that prompted me to write down the book that is right now, in your hands.

In case the result is a single way solution then?

The universe that is, right now, before us, would have been altogether a different one.

let's stand outside our milky way galaxy and let's try to analyse the mechanism of the universe. This mechanism is always result-oriented and the result is determined by the path- that is what has been named as, the sense. When I try to show the mechanism of the universe, outside of the galaxy- I belong to, then I could observe that although my age is very small still I am growing older. Now the question, that is right now, striking my mind is whether the same mechanism is followed here as is followed by the universe. The mechanism of my body made me understand a lot and the end product that is the result is because of those mechanisms. Now the very important question to be followed is whether a single mechanism gives rise to two different results?

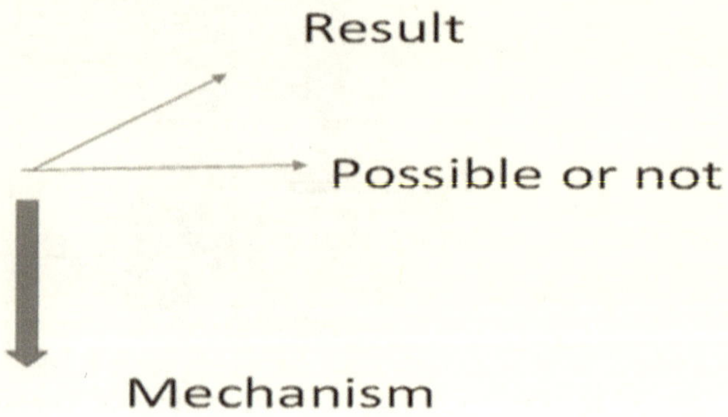

let's take a simple thing. Sir Faraday's observations. He has said," if the current passes through a conductor, it produces a magnetic field.

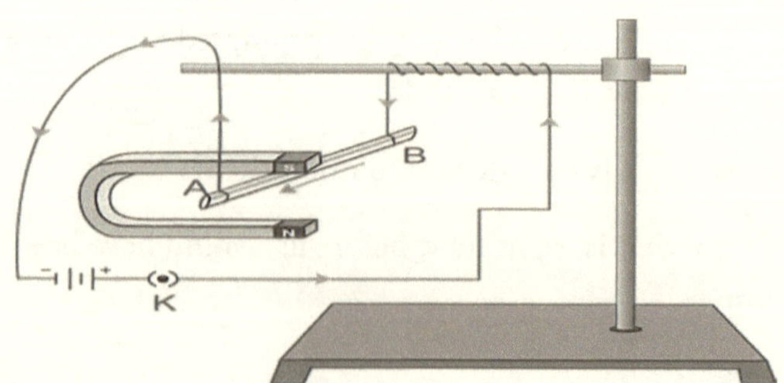

when I switched on, the current started flowing- I.e. electrons started moving.

this motion of electrons is the mechanism and the production of magnetic field or glowing of the lamp, is regarded as the result.

Now, let's talk of the Faradays electromagnetic induction

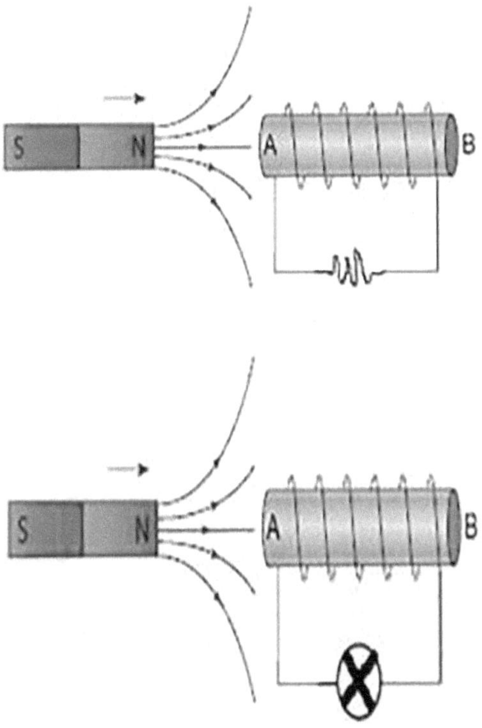

if there had been a single result or just one way mechanism, the universe would not have been so subtle, as it really exists. We could have found its ways and steps easily.

when I connected the Circuit, the mechanism for the bulb, resistance and speaker are different from one another. Now let's dig deep to understand all this. then we will check it out, whether two results are possible or not for the same mechanism.

Now, let's turn to dynamo and electric motor.

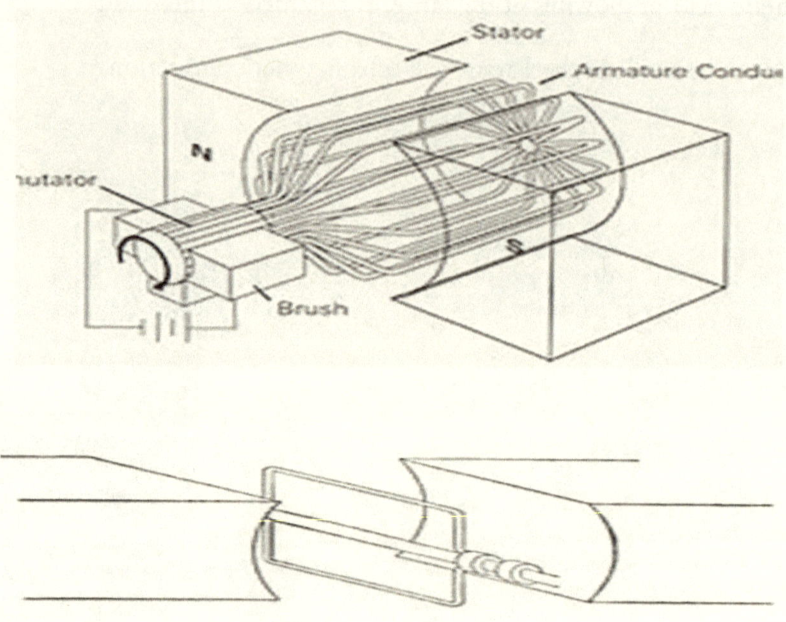

Case 1ˢᵗ

in case first, we move the magnetic field- and, you know changing magnetic field induces current.

Case 2ⁿᵈ

And in case second-and yes, you have got it right, that's a current carrying conductor produces magnetic field.

You would have found something similar at both the places. Did you? yes, you have felt right it is just the secret of magnetic field.

Now- let's think or pounce on to it a little more deeply

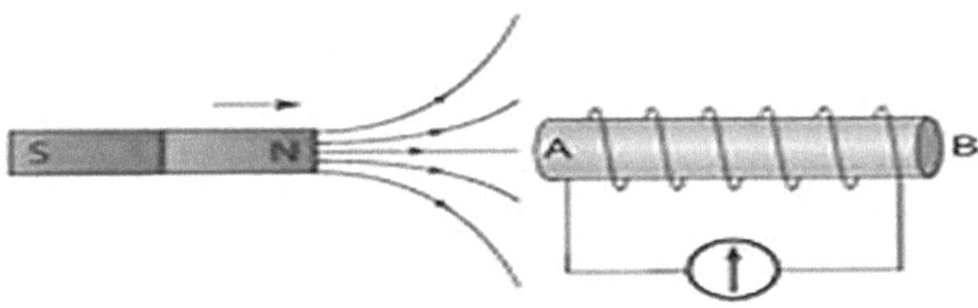

Can this moving magnetic field do something more? And if it can- that, too, without generating the current-

That is an all important question. Let's suspend a piece of iron by means of a fine thread,

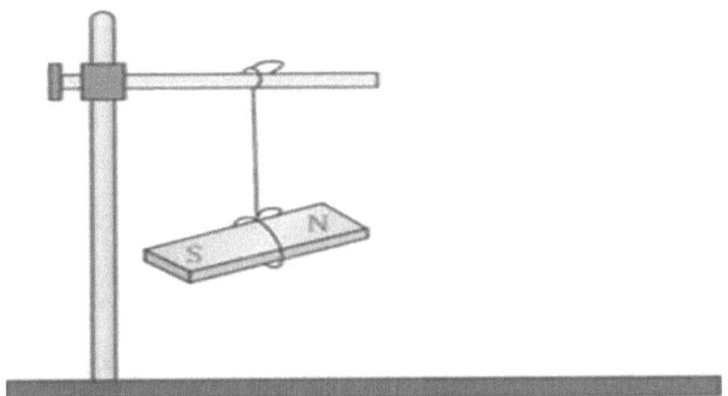

then move a magnet to and fro towards the piece of iron, we will see the iron piece starts the "to and fro" movement. It suggests that we kept the mechanism same that is changing magnetic field. At one instance, the ultimate product was that of current and at this instance, we kept two different situations and it is because of this that the changing situations have provided us with the different results and it is for the same reason, and is quite evident that the situations have made the universe all the more complicated. it is now evident that the results could vary and can be more than one. Now, we can infer from all this that when the universe began its course, the situations could have been fewer and, I say, fewer only, but as it continued to exist with its course of time, the ratio of the situations increased in number or quantity that gave the universe a reason and a chance, too, to became subtle, otherwise if the situations wouldn't have existed or increased, it would have been quite simpler in form and in a straight way and we could have imagined and understood each and every step, each and every phenomenon there of it- this is the 2nd time, I am mentioning this point over here,

Now the question arises-

What is after all the sense of physics and how does it work?

let's simplify it in a sort of figure a flow chart we will first look into two things- the process and the results.

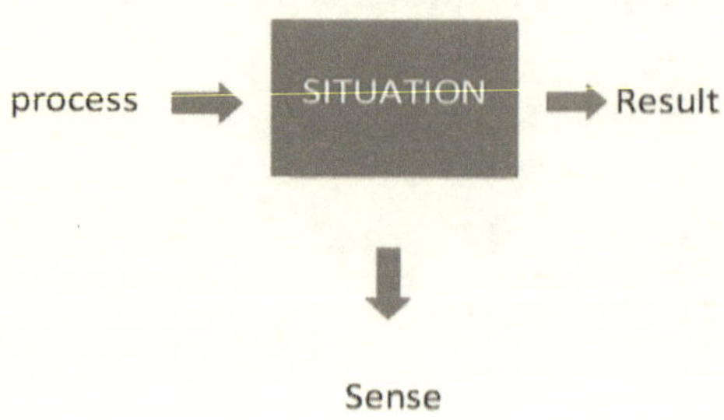

the things that determine the process are – nature, texture, properties, and mechanism which involve in or during a process. Now the question is regarding what this that is?

it refers to energy, mass and time

Now, our next figure would be like

To understood all this, let's isolate a light ray- there arise two situations to it.

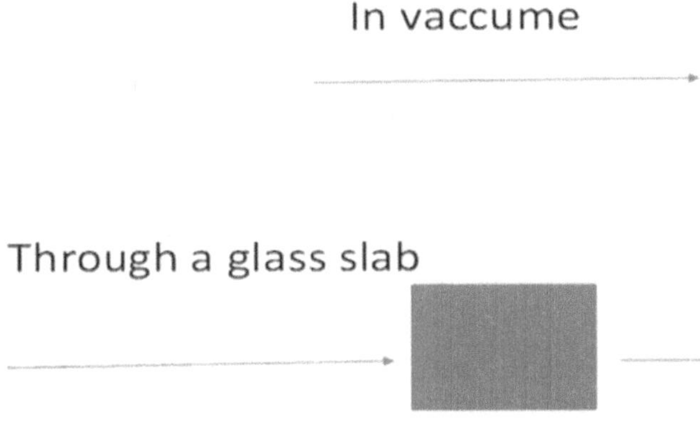

we can easily observe that the speed of light changes [decrease] when a ray of light pass through a glass slab- why? This is our result. Here

Now let's try to analyse a bit deeply, this phenomenon, whatever situations we provide for the light ray to pass, which one gave us a different result- Obviously, the glass slab. In fact, the glass slab had already predicted the future course of the light ray. it had already sensed how it's texture would behave with the ray of light as it passes through it. And this power of prediction or prediction the future course of anything may be, well, regarded as the sense of physics.

I am so confident that I consider when the universe started its journey, my existence was certain my existence, is not and cannot be, in any case, a mere probability or just a game of chance. Rather, a well predicted and well planned and well organised effort.

Do you know the words like "Yes" Or "No," You will definitely be-now let me detach you here into same aesthetics a personal, rather,-a very personal discourse- what happened is that a friend of mine asked me today if I would go for a picnic and I very confidently answered with the adverb of negative I.e. "No" by just claiming " no," brought me to an important discovery and it is- yes you got it right- this is not a mere word rather the prediction of the future. Had the universe ever done the same thing, it wouldn't have been so subtle, so complicated- rather the very simple one and it is what it is, only because compelled by these processes, situations and the results out of those arising situations hence the point.

Importance of pattern

Whatever information in the universe exists, it exists only because of these patterns. These patterns come of the building blocks of the universe. How do they combine? this all gets summed up to/in the memory of the universe. These patterns also do the task of sense making of the universe. Not only, this but the determining factor of the prediction also comes in these patterns earlier. This pattern gets genesis from the primary (1st) element of the universe- The one that came into being 1st. and it got complex day by day   I.e. it got through the evolution day by day- the first pattern of the universe was very basic in nature- and when two things collaborated in the universe, it got its pattern complexed a bit.

I am in the search of that very pattern. The pattern is additive in nature. when "pattern A" and "pattern B" gets combined together a new pattern C gets existed that resembles in properties with A and B patterns that is C = A + B.

The pattern is subtractive in nature. It means when two patterns that is "pattern A" and "pattern B" get combined they give rise to a third pattern C which is altogether different from the previous patterns that is A and B that is C is neither A nor B. can the patterns cancel each other if yes, that, too, during their combination. The scientists named it "the matter" else "the antimatter".

Can it happen that two patterns came- closer by still they do not get combined. yes, sometimes pattern A and pattern B do not get combined despite coming very close by to each other this is done to the fact that they do not fulfil the basic conditions to get combined. Now the question arises, after all what are those conditions? This is something that you, as my reader, will get accustomed to, in my new book- and I will discuss it all these in that.

Do the patterns cancel the other patterns

Yes, when "pattern A" and "pattern B" come very close to each other. then one of these patterns tries to dominate fully the other and due to which the other pattern becomes recessive then its properties remain latent and do not come to the fore.

The sense of physics is not like our senses our sense is just the one that gives or provides us with the information I.e. our sense is information oriented. That, too related to the presence or absence. But the sense of universe/the sense of physics- it predicts the future. our senses, sometimes do wrong, get mislead or mislead us but the sense of physics never gets wrong. Is not and cannot ever be misled in prediction or predicting the things (the results).

The sense of physics is run or driven by the same pattern. Pattern, a sort of information, is what has given genesis to the present day complex-universe.

1st single pattern got formed in the universe when only a thing existed- then the sense came in-to being when some other thing took the genesis or existence in the universe.

And when it came in to being or got formed it started to do its own part [work].

So the sense is dependent on the pattern.

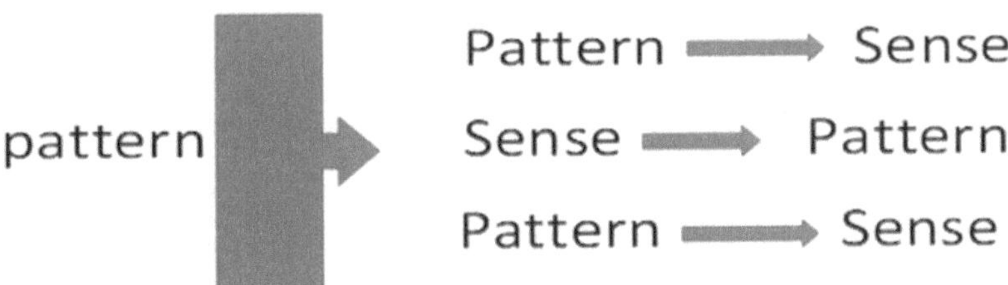

Can a sense becomes a pattern

 yes it can happen and there is every opportunity of the sense becoming a pattern and vice versa. you would have understood, why we human beings are quite unable to recognise and realise the universe- this is because we do not have felt or understood any of the processes going in the universe with any of our senses or sense organs- we have just tried to collect a little bit information and tried to make or form a picturesque out of it. Formed and formulated a view of it - that is why what-else happens in the universe we just go on wagging our tongue to give best possible explanation to it and nothing else- we only try to make a view that this or that might have happened but are quite unable to feel/make others feels that this or that has technically or practically happened and we are just unable to tell why that has happened.

Gold foil experiment

We have created a model for detecting the position or location of the protones (nucleus) of the atom. or how does an atom look like the laws of thermodynamics gave us the way and vision to analyse the process.

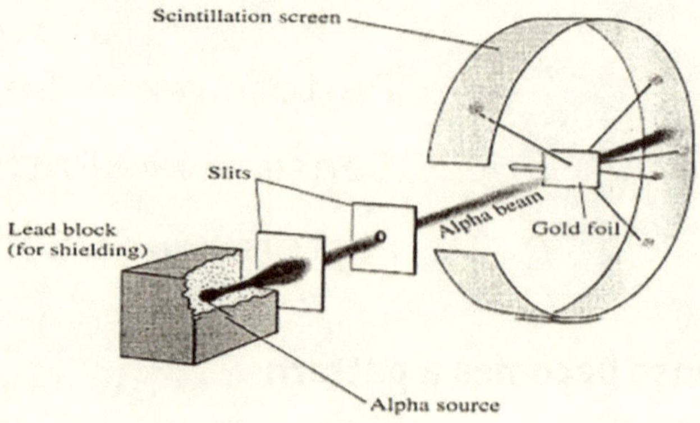

1st law:.

Almost all the processes, when got analysed we reached to the conclusion that:

Input energy = output energy

As then, we formulated the law that energy can (neither) not be created nor destroyed (by us). A question bigger than that is how universe came into being out of nothing and another related question is what does nothing connote in physics? we know that heat is a form of energy that gives us sensation of warmth.

Now the question arises, what else we feel, does the behaviour of the heat exists so? Our skin felt the heat, and our mind took the final decision in perceiving what or how it looked like (felt). Now the question arises- what else our mind thought it to be, does it (heat) actually exists as such? Or does heat show the same or similar behaviour as our mind thinks it to be? This is the question that is put for you to answer? Yes- You are right. it is obviously "no"- different species observe the same situation differently due to the basic (patterns) differences lying in their senses. A simple question would be like: if we tried to split any of the fundamental quantities that what is it that we regarded as the quantity? That's perfectly the thing that we call as nothing. Now, does that nothing holds any /same sense? Yes- indeed there does exist a sense within every nothing or nothingness that is pertaining to making up of a thing.

We go on splitting the thing (quantity) further till we reach that nothing or nothingness I.e. ΔP approach to nothing.

ΔP 0

When and if we reach to nothing or nothingness of anything, we will meet a thing there that we regard as information that information exists in terms of a mathematical equation and is quite underlying principle of universe.

So, this sense exists in the form of information.

I,I + I,I = I,I,I,I

I,I,I × I,I = I,I,I,I,I,I

I,I,I,I,I,I ÷ I,I = I,I,I,

I,I,I,I,I,I,I,I,I,I, - I,I, = I,I,I,I,I,I,I,I,

Yes, you're right the universe follows simple mathematics

And this sense in the form, of information, generally exists in the form of +,-,×,÷ etc. The "+sign" shows and detects how various quantities combine together.

The –sign detects and determines how will some or several quantities get eradicated in future.

The ÷ sign infers the division of the various quantities, the × Sign represents the multiplicity of various quantities. so this sense bearing the shape of these mathematical observation exists as pure mathematical equations. whenever any equation comes into being, a huge effect of it gets produced on to the universe.

How does information gets pattern.

The pattern is possessive of the information. it's the sole property of space to create a pattern. The pattern of the universe is made of that building elements that we call

¯_(ツ)_/¯.

The way, binary codes get stored in a computer, the same way, the information in the form of mathematical equation gets stored and collected in the universe. this information exists in the universe quite in the simple form. so the universe by itself is just simple but due to our approach, it is more complicated. it is because we are bound to predict some/all of its phenomenon and due to that prediction, we think it to be complex to understand it and this brings us to the fact that this addition, subtraction, multiplication, or division based its nature make it all the more simple- and this being simple gives it the increased effect towards finding what would happen to the universe in the next other phase.? What has actually happened to it previously? Whatever happens in the universe or whatever happened till date or what else is going to happen next, this all is stored up in its memory and with it. It can already predict what will /would happen the next.

What sense can do?

It is the sense that determines the future. And it is because of it, that present exists and whatever has happened in the past, it is because of it.

Whatever exists in the universe, does it know or does it perceive what else is going with it or what else might happen with it?

Science of flame

Structure of flame on zero gravity and on the earth.

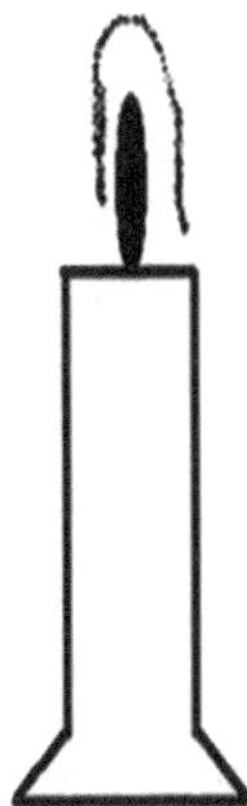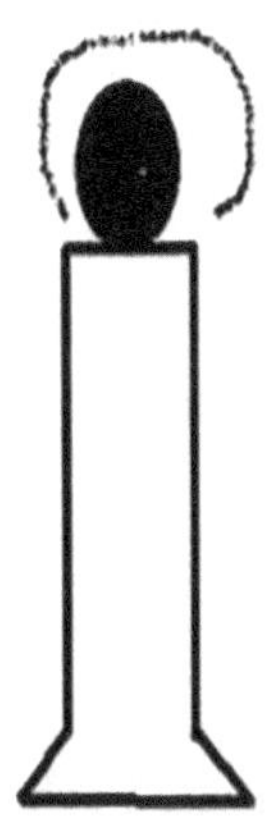

The structure of flame given here you can treat it the result and what is the cause behind this. it is the same thing that determines the results the concept of gravity has a greater role behind it all. this is because we provided the flame all the necessary conditions on zero gravity.

It means that gravity is responsible for the structure of the flame it suggests that the outcome of the gravity is to attract the things- and this effects of attraction has a greater cause on the structure of the flame and how does it affect its structure, this is the sense.

Now the question arises whether this is the compulsion or choice (licence).

This is evidently a choice, a licence but this licence is created out of the information

I want to tell you only this thing that this all has been done because of the absence or presence of something. The absence of something or presence of something has shown us the result. That absence or presence of something is either the internal or the external factor of its process- and these are the very ways that tend to bring to us to the said results. Or it could be the effect of those external factors in the process. And as far as i think, I consider this thing of utmost importance with respect to the best logic behind the solution. I don't want to confuse you rather, I want to tell you that how much important the effect of external factors is in deceiving a process- and we can only deceive the senses- and the sole aim of this book is evident from its name-

That is the external factors don't leave anything in deceiving and deceiting the outcome of the process.

The senses can't be treated as advanced on the mere potential of the predictions it can reveal- because the results can't be certain always. These results often vary. It is the very capacity of the varying results that compels us to think that the senses of the universe are not that much advanced yet.

Importance of sequence in the universe

The pattern of the universe consists of the sequence and this sequence is the driving force to determine the quantities of the universe. Different elementary particles got genesis only due to the sequence- and this sequence is the framework of dimensions.

If we succeed in breaking the quark,then we shall arrive at the rubber-band type structure a quantity and as we go on further dividing the thing, almost to the level zero, then we get sequence. This is the sequence that creates the pattern, I fear, that this sequence should not be anybody's dream- and if it so happens, then we are a mere part of a single dream.

And in case it does not happen so, then there is a master plan behind the universe and this master plan is in the form of information and it exists in a pattern that has a sequence- and this sequence and pattern is the backbone of the universe- that is present before us and it is the same thing that shaped it in the past and it will be thus in the future as well- what else happened or happens or will happen, it is all due to this pattern. And if we talk of the universe at the pattern level then "god equation" seems to be possible but if we talk of the sequence it seems dead impossible.

I was busy in listening to some beautiful number (song) over the cell phone that, too, in a meditating way, eyes closed. My papa took away my phone and I could perceive the cell phone distancing from me due to the fall in sound. So, this fall in sound gave me an idea about the location of my cell phone.

Now there exists a condition one.

If papa would have increased the volume of the cell phone while keeping it away from me, could I have still detected the location?.

No answer to this? So we need to make use of another sense to detect this. That is obviously, the sense of sight.

Now, the question arises, does physics (science) possess only a single sense or multi senses.

And the answer to this question is yes, there exist different senses in physics predictions of processes of universe. Does process of universe predict the future?

And what happens the next? If the behaviour preceded ever in the universe, it preceded the mechanism. Whenever (any) thing came into being, the first thing that existed ever, it is the behaviour thereof. We can understand the behaviour a bit good. It is the prediction that is regarded as the sense in/of physics. if the universe predicted the future, it is then the sense, The sense of physics. Evolution, almost, transcended later this mechanism whether it be the evolution of the universe or the biological one lets, then, try to understand it a bit deeply by understanding the principles of the helicopter or aeroplane.

Process   ⟶   behaviour

Process of aeroplane

First the oil burns /combustion.

The router moves which, in turn, pushes down the air that results in the take-off of the aeroplane [helicopter]. During this process, the result we get, has a definite beginning and the ultimate end. what else the result we get, it is just the prediction. We call the same thing as sense in universe.

When this universe came into being, it decided then end there what is going to happen in the future and what is it that would not happen. We have just discussed how the helicopter works. The behaviour that we observed, it tells us that first the sense got its existence that we call as the prediction then behaviour.

If there had not been any prediction in the universe, the universe would have evolved through impossible many ways. the existence of predictions was possible during the very initial phase of the universe.

What else happens to the universe, it is the same sense that determines it. In fact, the behaviour usually comes first but it is when the process would have started but the observers do not observe it till the process gets completed.

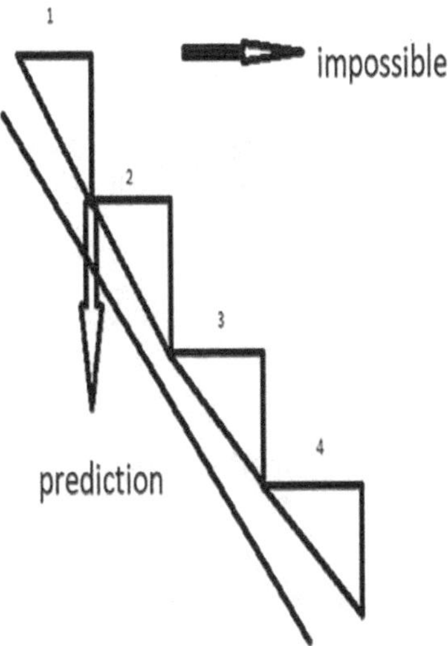

The valves get predicted earlier and the amount of effect,too, gets predicted. This is because the quantization of charge rule provides us the list of valves earlier.

The Sense of Physics

How much Vs how? The sense of physics gives us a clear cut view and the sense tells us how? the sense tells us the predictions of the future.

Now the question to be asked is how much advanced the sense should be?

There are, undoubtedly, the drawbacks of very advanced sense. So, the sense should not be too much advanced. let's talk of our own sense of sight.

If It be too advanced that we can see even the microscopic entities like viruses and bacteria would our life be possible?

No, the life would have been impossible because we would have remained confined and concerned about these entities and our life would have been impossible as such.

Now the sense of physics, it is not that much advanced as of ours- had it been advanced enough, the activities that are quite impossible in the universe, would not have turned possible. it is the phenomenon of not being too advanced that gives rise to the impossible. Now the question exists, has the universe followed the impossible ways?

And yes, if it has, it would have taken a new shape in existence.

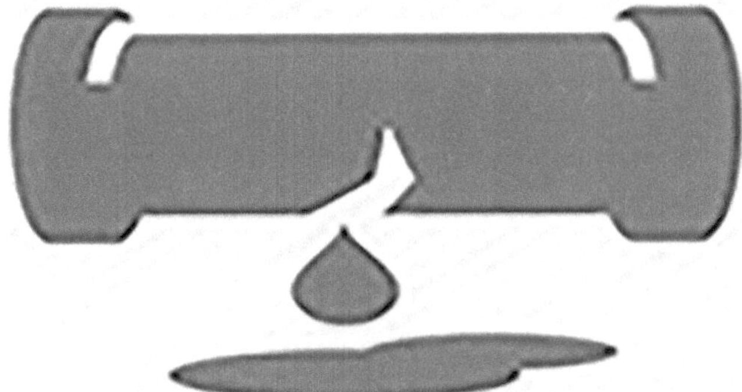

You see, this pipe has a drawback. It leaks- Now if we give it the shape of a shower- Let's see what happens.

Marvellous! This art worked for us.

Does universe,too, takes the art into consideration ever. The advancement of sense gives rise to the "impossible" in the universe. And that is why that prompted Richard dawkin's, to write /publish the book the "blind watch maker"

Does the universe follow the impossible ways. the answer to this question is "yes". It does and it's true. It takes the art into consideration- that is how the universe has a clear beginning and the ultimate end.

+ 1=2

2 + 1=3

3 + 2=5

4 + 4=8

This list that is been prepared here it is already been there,

It is the future prediction table. It's already been there quite existing. let's talk of an another table

1.1 + 0.9 =2

2.2 + 1.8 =4

3.3 + 0.7 =4

This result that we have reached at, is just the outcome of logic.

This is because the logic has occupied all the possibilities keeping the sense oriented path quite open. And that path gives rise to the process that advances forth

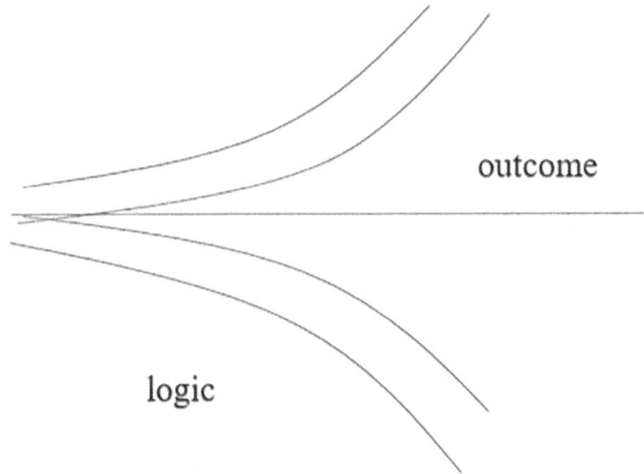

Sense is a step by step process

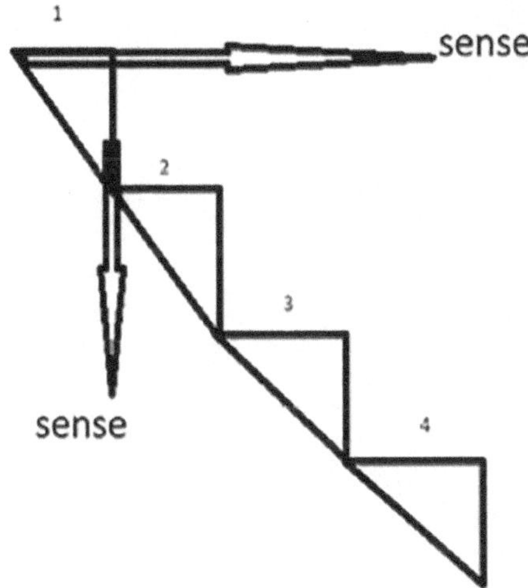

from beginning up to the end, that is ultimate and inevitable evidences that the law of conversation of mass and energy is the results of the art. if there had not been this art then the possibility of the ultimate end of the universe would have become impossible because the law of conservation of mass and energy signifies that there is an end that is inevitable. Now, is this the thing that have been allowed by the sense of physics? yes, I have already maintained that end that system of sense is not that much advanced yet this is just the type of the sense that helps in predicting the things and not of the type that provides 'certainty'. As the matter of fact, the prediction can change subject to the change of situation or environment. And in the same fashion, if or whenever the texture, behaviour or properties get changed, the results change.

The gravitational pull in the black hole is that much great that even the fast moving light cannot escape. How does this property arise? or is this the result of the addition of the masses or some other reason is there behind it? The addition takes us to the realm of the sense of logic. The result, all the more, is based on the logic

$$1+1=2$$

Is this the logic or is 2 the result of 1+1 and number 2 can also be the results of 0.9 and 1.1

$$0.9+1.1=2$$

The results can be same even if the sense can be varying- the sense (s) that gives rise to a definite process

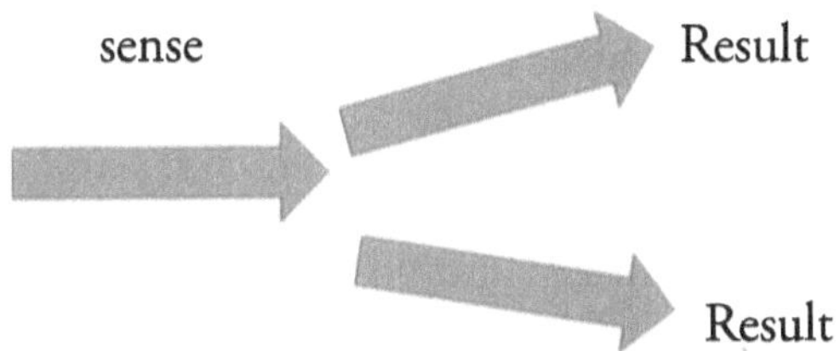

It is like someone is showing or leading a blind man to some path that is why I regard these senses as blind ones. the results get predicted of the universe but the initial quantity should not change and all those factors that have a direct bearing on the results should not vary either. we have already illustrated the example

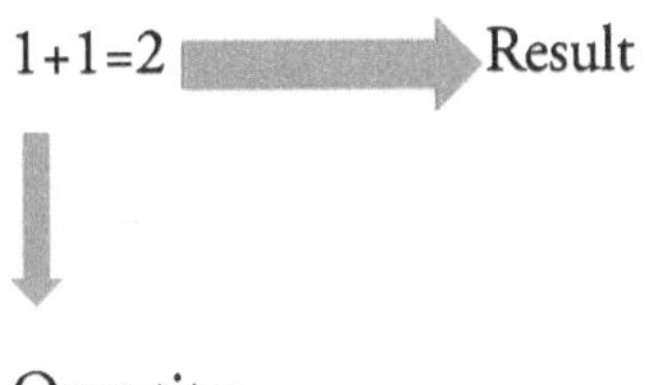

Quantity

Does the result become possible through a solitary way? or can there exists some other ways to lead to some results? Yes, there can be many other ways 0.9 +1.1=2 as this. All is possible in universe.

Sense does permit it but in universe there exists another thing that does not allow it and that is the logic of universe.

The sense can err sometimes yes- it could have but the logic did not permit it.

Let's move to the law of quantization of charge. Let's talk of them according to the quantization of charge, the future can be predicted...... it gives certain valves 1,2,3,4,5,6,7,8, 9.etc

These values are determined by the law of quantization of the charges and it is the very thing that turn out to be the sense- that in turn, provides us certain definite values...

I am mindful of the facts that various of your queries have remained unaddressed in this publication. This is predominantly because I want to create a creative-learning atmosphere for you. So that you will adopted a new way of thinking in analysing the universe we are living in. A promise, I shall leave no space unnoticed and unaddressed in my next publications. So be patient and have fun...